A Covnter Blaste to Tobacco
& Daemonologie

by
King James I of England

Printed on acid free ANSI archival quality paper.

© 2011 Benediction Classics, Oxford.

Contents

A Covnter Blaste to Tobacco	1
Preface	2
A Covnter-Blaste to Tobacco	5
Daemonologie	17
Preface	18
First Booke	21
Chapter 1	22
Chapter 2	26
Chapter 3	28
Chapter 4	30
Chapter 5	33
Chapter 6	36
Chapter 7	40
Seconde Booke	43
Chapter 1	44
Chapter 2	47
Chapter 3	49
Chapter 4	52
Chapter 5	56
Chapter 6	60
Chapter 7	63
Thirde Booke	65
Chapter 1	66
Chapter 2	70
Chapter 3	73
Chapter 4	76
Chapter 5	79
Chapter 6	82
Newes from Scotland	86
To the Reader	87
Discourse	88

A Covnter Blaste to Tobacco

by
King James I of England

Edited by
Edmund Goldsmid, F.R.H.S.
Edinburgh, 1884

Preface

As euery humane body (deare Countrey men) how wholesome soeuer, be notwithstanding subiect, or at least naturally inclined to some sorts of diseases, or infirmities: so is there no Common-wealth, or Body-politicke, how well gouerned, or peaceable soeuer it bee, that lackes the owne popular errors, and naturally enclined corruptions: and therefore is it no wonder, although this our Countrey and Common-wealth, though peaceable, though wealthy, though long flourishing in both, be amongst the rest, subiect to the owne naturall infirmities. We are of all Nations the people most louing and most reuerently obedient to our Prince, yet are wee (as time has often borne witnesse) too easie to be seduced to make Rebellion, vpon very slight grounds. Our fortunate and off prooued valour in warres abroad, our heartie and reuerent obedience to our Princes at home, hath bred vs a long, and a thrice happy peace: Our Peace hath bred wealth: And Peace and wealth hath brought foorth a generall sluggishnesse, which makes vs wallow in all sorts of idle delights, and soft delicacies, The first seedes of the subuersion of all great Monarchies. Our Cleargie are become negligent and lazie, our Nobilitie and Gentrie prodigall, and solde to their priuate delights, Our Lawyers couetous, our Common-people prodigall and curious; and generally all sorts of people more carefull for their priuate ends, then for their mother the Common-wealth. For remedie whereof, it is the Kings (as the proper Phisician of his Politicke-body) to purge it of all those diseases, by Medicines meete for the same: as by a certaine milde, and yet iust form of gouernment, to maintaine the Publicke quietnesse, and preuent all occasions of Commotion: by the example of his owne Person and Court, to make vs all ashamed of our sluggish delicacie, and to stirre vs up to the practise againe of all honest exercises, and Martiall shadowes of VVarre; As likewise by his, and his Courts moderatenesse in Apparell, to make vs ashamed of our prodigalitie: By his quicke admonitions and carefull overseeing of the Cleargie to waken them vp againe, to be more diligent in their Offices: By the sharpe triall, and seuere punishment of the partiall, couetous and bribing Lawyers, to reforme their corruptions: And generally by the example of his owne Person, and by the due execution of good Lawes, to reform and abolish, piece and piece, these old and euill grounded abuses. For this will not bee Opus vnius diei, but as euery one of these diseases, must from the King receiue the owne cure proper for it, so are there some sorts of abuses in Common-wealths, that though they be of so base and contemptible a condition, as they are too low for the Law to looke on, and too meane for a King to interpone his authoritie, or bend his eye vpon: yet are they corruptions, as well as the greatest of them. So is an Ant an Animal, as well as an Elephant: so is a VVrenne Auis, as well as a Swanne, and so is a small dint of the Toothake, a disease as well as the fearefull Plague is. But for these base sorts of corruption in Common-wealthes, not onely the King, or any inferior Magistrate, but Quilibet è populo may serue to be a Phisician, by discouering and impugning the error, and by perswading reformation thereof.

And surely in my opinion, there cannot be a more base, and yet hurtfull corruption in a Countrey, then is the vile vse (or other abuse) of taking Tobacco in this Kingdome, which hath moued me, shortly to discouer the abuses thereof in this following little Pamphlet.

If any thinke it a light Argument, so it is but a toy that is bestowed upon it. And since the Subiect is but of Smoke, I thinke the fume of an idle braine, may serue for a sufficient battery against so fumous and feeble an enemy. If my grounds be found true, it is all I looke for; but if they cary the force of perswasion with them, it is all I can wish, and more than I can expect. My onely care is, that you, my deare Countrey-men, may rightly conceiue euen by this smallest trifle, of the sinceritie of my meaning in great matters, never to spare any paine that may tend to the procuring of your weale and prosperitie.

A Covnter-Blaste to Tobacco

That the manifolde abuses of this vile custome of *Tobacco* taking, may the better be espied, it is fit, that first you enter into consideration both of the first originall thereof, and likewise of the reasons of the first entry thereof into this Countrey. For certainely as such customes, that haue their first institution either from a godly, necessary, or honorable ground, and are first brought in, by the meanes of some worthy, vertuous, and great Personage, are euer, and most iustly, holden in great and reuerent estimation and account, by all wise, vertuous, and temperate spirits: So should it by the contrary, iustly bring a great disgrace into that sort of customes, which hauing their originall from base corruption and barbarity, doe in like sort, make their first entry into a Countrey, by an inconsiderate and childish affectation of Noueltie, as is the true case of the first inuention of *Tobacco* taking, and of the first entry thereof among vs. For *Tobacco* being a common herbe, which (though vnder diuers names) growes almost eueryewhere, was first found out by some of the barbarous *Indians*, to be a Preseruatiue, or Antidot against the Pockes, a filthy disease, whereunto these barbarous people are (as all men know) very much subiect, what through the vncleanly and adust constitution of their bodies, and what through the intemperate heate of their Climate: so that as from them was first brought into Christendome, that most detestable disease, so from them likewise was brought this vse of *Tobacco*, as a stinking and vnsauorie Antidot, for so corrupted and execrable a Maladie, the stinking Suffumigation whereof they yet vse against that disease, making so one canker or venime to eate out another.

And now good Countrey men let vs (I pray you) consider, what honour or policie can mooue vs to imitate the barbarous and beastly maners of the wilde, godlesse, and slauish *Indians*, especially in so vile and stinking a custome? Shall wee disdaine to imitate the maners of our neighbour *France* (hauing the stile of the first Christian Kingdom) and that cannot endure the spirit of the Spaniards (their

King being now comparable in largenes of Dominions to the great Emperor of *Turkie*). Shall wee, I say, that haue bene so long ciuill and wealthy in Peace, famous and inuincible in Warre, fortunate in both, we that haue bene euer able to aide any of our neighbours (but neuer deafed any of their eares with any of our supplications for assistance) shall we, I say, without blushing, abase our selues so farre, as to imitate these beastly *Indians*, slaues to the *Spaniards*, refuse to the world, and as yet aliens from the holy Couenant of God? Why doe we not as well imitate them in walking naked as they doe? in preferring glasses, feathers, and such toyes, to golde and precious stones, as they do? yea why do we not denie God and adore the Deuill, as they doe?[1]

Now to the corrupted basenesse of the first vse of this *Tobacco*, doeth very well agree the foolish and groundlesse first entry thereof into this Kingdome. It is not so long since the first entry of this abuse amongst vs here, as this present age cannot yet very well remember, both the first Author,[2] and the forme of the first introduction of it amongst vs. It was neither brought in by King, great Conquerour, nor learned Doctor of Phisicke.

With the report of a great discouery for a Conquest, some two or three Sauage men, were brought in, together with this Sauage custome. But the pitie is, the poore wilde barbarous men died, but that vile barbarous custome is yet aliue,[3] yea in fresh vigor: so as it seemes a miracle to me, how a custome springing from so vile a ground, and brought in by a father so generally hated, should be welcomed vpon so slender a warrant. For if they that first put it in practise heere, had remembred for what respect it was vsed by them from whence it came, I

---

[1] This argument is merely that because an inferior race has made a discovery, a superior one would be debasing itself by making use of it.

[2] By Sir Walter Raleigh, one of the greatest and most learned men of the age, whose head the author cut off, partly influenced, no doubt, by his detestation of tobacco. Smokers may therefore look upon the author of the "History of the World" as the first martyr in their cause.

[3] A centenarian has recently died, the papers relate, who, till within a few days of his death, was in perfect health, having been a constant smoker, but was unfortunately induced by his friends to give up the habit, from which moment he rapidly sank. Probably these barbarians were affected in the same manner.

am sure they would haue bene loath, to haue taken so farre the imputation of that disease vpon them as they did, by vsing the cure thereof. For *Sanis non est opus medico*, and counter-poisons are neuer vsed, but where poyson is thought to precede.

But since it is true, that diuers customes slightly grounded, and with no better warrant entred in a Commonwealth, may yet in the vse of them thereafter, prooue both necessary and profitable; it is therefore next to be examined, if there be not a full Sympathie and true Proportion, betweene the base ground and foolish entrie, and the loathsome, and hurtfull vse of this stinking Antidote.

I am now therefore heartily to pray you to consider, first vpon what false and erroneous grounds you haue first built the generall good liking thereof; and next, what sinnes towards God, and foolish vanities before the world you commit, in the detestable vse of it.[4]

As for these deceitfull grounds, that haue specially mooued you to take a good and great conceit thereof, I shall content myselfe to examine here onely foure of the principals of them; two founded vpon the Theoricke of a deceiuable apparance of Reason, and two of them vpon the mistaken Practicke of generall Experience.

First, it is thought by you a sure Aphorisme in the Physickes, That the braines of all men, being naturally colde and wet, all dry and hote things should be good for them; of which nature this stinking suffumigation is, and therefore of good vse to them. Of this Argument, both the Proposition and Assumption are false, and so the Conclusion cannot but be voyd of it selfe. For as to the Proposition, That because the braines are colde and moist, therefore things that are hote and drie are best for them, it is an inept consequence: For man beeing compounded of the foure Complexions (whose fathers are the foure Elements) although there be a mixture of them all in all the parts of his body, yet must the diuers parts of our *Microcosme* or little world within ourselves, be diuersly more inclined, some to one, some to another complexion, according to the diuersitie of their vses, that of these discords a perfect harmonie may bee made vp for the maintenance of the whole body.

The application then of a thing of a contrary nature, to any of these parts is to interrupt them of their due function, and by consequence hurtfull to the health of the whole body. As if a man, because

---

[4] Had the royal pedant ever heard of locking the stable door after the horse has been stolen?

the Liuer is hote (as the fountaine of blood) and as it were an ouen to the stomache, would therefore apply and weare close vpon his Liuer and stomache a cake of lead; he might within a very short time (I hope) be susteined very good cheape at an Ordinairie, beside the cleering of his conscience from that deadly sinne of gluttonie. And as if, because the Heart is full of vitall spirits, and in perpetuall motion, a man would therefore lay a heauy pound stone on his breast, for staying and holding downe that wanton palpitation, I doubt not but his breast would bee more bruised with the weight thereof, then the heart would be comforted with such a disagreeable and contrarious cure. And euen so is it with the Braines. For if a man, because the Braines are colde and humide, would therefore vse inwardly by smells, or ontwardly by application, things of hot and drie qualitie, all the gaine that he could make thereof would onely be to put himselfe in a great forwardnesse for running mad, by ouer-watching himselfe, the coldnesse and moistnesse of our braine beeing the onely ordinarie meanes that procure our sleepe and rest. Indeed I do not denie, but when it falls out that any of these, or any part of our bodie growes to be distempered, and to tend to an extremetie, beyond the compasse of Natures temperate mixture, that in that case cures of contrary qualities, to the intemperate inclination of that part, being wisely prepared and discreetely ministered, may be both necessarie and helpfull for strengthning and assisting Nature in the expulsion of her enemies: for this is the true definition of all profitable Physicke.

But first these Cures ought not to bee vsed, but where there is neede of them, the contrarie where of, is daily practised in this generall vse of *Tobacco* by all sorts and complexions of people.

And next, I deny the minor of this argument, as I haue already said, in regard that this *Tobacco*, is not simply of a hot and dry qualitie; but rather hath a certaine venemous facultie ioyned with the heate thereof, which makes it haue an Antipathie against nature, as by the hatefull smell thereof doeth well appeare. For the nose being the proper Organ and convoy of the sense of smelling to the braines, which are the onely fountaine of that sense, doeth euer serue vs for an infallible witnesse, whether that Odour which we smell, be healthfull or hurtfull to the braine (except when it fals out that the sense it selfe is corrupted and abused through some infirmitie, and distemper in the braine.) And that the suffumigation thereof cannot haue a drying qualitie, it needes no further probation, then that it is a smoake, all smoake and vapour, being of it selfe humide, as drawing neere to the nature of the ayre, and easie to be resolued againe into water, whereof there needes no other proofe but the meteors, which being bred of nothing else but of the vapours and exhalations sucked vp by the Sunne out of the earth, the Sea, and waters, yet are the same smoakie vapours turned, and transformed into Raynes, Snowes, Dewes, hoare Frostes, and such like waterie Meteors, as by the contrarie the raynie cloudes are often transformed and euaporated in blustering winds.

 The second Argument grounded on a show of reason is, That this filthie smoake, as well through the heat and strength thereof, as by a naturall force and qualitie, is able and fit to purge both the head and stomacke of Rhewmes and distillations, as experience teacheth, by the spitting and auoyding fleame, immeadiately after the taking of it. But the fallacie of this Argument may easily appeare, by my late preceding description of the Meteors. For euen as the smoakie vapours sucked vp by the Sunne, and staied in the lowest and colde Region of the ayre, are there contracted into Cloudes and turned into raine and such other watery Meteors: So this stinking smoake being sucked vp by the Nose, and imprisoned in the colde and moyst braines, is by their colde and wett facultie, turned and cast foorth againe in waterie distillations, and so are you made free and purged of nothing, but that wherewith you wilfully burdened yourselues: and therefore are you no wiser in taking *Tobacco* for purging you of distillations, then if for preuenting the Cholike you would take all kinde of windie meates and drinkes, and for preuenting the Stone, you would take all kinde of meates and drinkes, that would breede grauell in the Kidneys, and then when you were forced to auoyde much winde out of your stomacke, and much grauell in your Vrine, that you should attribute the thanke thereof to such nourishments as bred those within you, that behoued either to be expelled by the force of nature, or you to haue *burst at the broad side*, as the Prouerbe is.

 As for the other two reasons founded vpon experience. The first of which is that the whole people would not haue taken so generall a good liking thereof, if they had not by experience found it verie soueraigne, and good for them: For answere thereunto how easily the mindes of any people, wherewith God hath replenished this world, may be drawn to the foolish affectation of any noueltie, I leaue it to the discreet iudgement of any man that is reasonable.

 Doe we not dayly see, that a man can no sooner bring ouer from beyond the Seas any new forme of apparell, but that hee cannot bee thought a man of spirit, that would not presently imitate the same? And so from hand to hand it spreades, till it be practised by all, not for any commoditie that is in it, but only because it is come to be the fashion. For such is the force of that naturall Selfe-loue in euery one of vs, and such is the corruption of enuie bred in the brest of euery one, as we cannot be content vnlesse we imitate euerything that our fellowes doe, and so prooue our selues capable of euerything whereof they are capable, like Apes, counterfeiting the maners of others, to our owne destruction.[5] For let one or two of the greatest Masters of Mathematickes in any of the two famous Vniuersities, but constantly affirme any cleare day, that they see some strange apparition in the skies: they will I warrant you be seconded by the greatest part of the Students in that profession: So loath will they be, to bee thought inferiour to their fellowes, either in depth of knowledge or sharpnesse of sight: And therefore the generall good liking and imbracing of this foolish custome, doeth but onely proceede from that affectation of noueltie, and popular errour, whereof I haue already spoken.[6]

The other argument drawen from a mistaken experience, is but the more particular probation of this generall, because it is alleaged to be found true by proofe, that by the taking of *Tobacco* diuers and very many doe finde themselves cured of diuers diseases as on the other part, no man euer receiued harme thereby. In this argument there is first a great mistaking and next a monstrous absurditie. For is it not a very great mistaking, to take *Non causam pro causa*, as they say in the Logicks? because peraduenture when a sicke man hath had his disease at the height, hee hath at that instant taken *Tobacco*, and afterward his disease taking the naturall course of declining, and consequently the patient of recouering his health, O then the *Tobacco* forsooth, was the worker of that miracle. Beside that, it is a thing well knowen to all Physicians, that the apprehension and conceit of the patient hath by wakening and vniting the vitall spirits, and so strengthening nature, a great power and vertue, to cure diuers diseases. For an euident proofe of mistaking in the like case, I pray you what foolish boy, what sillie wench, what olde doting wife, or ignorant countrey clowne, is not a

---

[5] The previous arguments can of course have no weight in our day, but this tendency to imitate others is as true now as then. Evidently, if the Darwinian theory holds good, a matter of three centuries is not sufficient to cause any perceptible diminution in the strength of original instinct inherited from the ape.

[6] Time has taken upon itself to upset this argument; for though the novelty may certainly be said to have worn off, the habit itself is more firmly rooted than ever.

Physician for the toothach, for the cholicke, and diuers such common diseases? Yea, will not euery man you meete withal, teach you a sundry cure for the same, and sweare by that meane either himselfe, or some of his neerest kinsmen and friends was cured? And yet I hope no man is so foolish as to beleue them. And al these toyes do only proceed from the mistaking *Non causam pro causa*, as I haue already sayd, and so if a man chance to recouer one of any disease, after he hath taken *Tobacco*, that must haue the thankes of all. But by the contrary, if a man smoke himselfe to death with it (and many haue done) O then some other disease must beare the blame for that fault. So do olde harlots thanke their harlotrie for their many yeeres, that custome being healthfull (say they) *ad purgandos Renes*, but neuer haue minde how many die of the Pockes in the flower of their youth. And so doe olde drunkards thinke they prolong their dayes, by their swinelike diet, but neuer remember howe many die drowned in drinke before they be halfe olde.

And what greater absurditie can there bee, then to say that one cure shall serue for diuers, nay, contrarious sortes of diseases? It is an vndoubted ground among all Physicians, that there is almost no sort either of nourishment or medicine, that hath not some thing in it disagreeable to some part of mans bodie, because, as I haue already sayd, the nature of the temperature of euery part, is so different from another, that according to the olde prouerbe, That which is good for the head, is euill for the necke and the shoulders. For euen as a strong enemie, that inuades a towne or fortresse, although in his siege thereof, he do belaie and compasse it round about, yet he makes his breach and entrie, at some one or few special parts thereof, which hee hath tried and found to bee weakest and least able to resist; so sicknesse doth make her particular assault, vpon such part or parts of our bodie, as are weakest and easiest to be ouercome by that sort of disease, which then doth assaile vs, although all the rest of the body by Sympathie feele it selfe, to be as it were belaied, and besieged by the affliction of that speciall part, the griefe and smart thereof being by the sense of feeling dispersed through all the rest of our members. And therefore the skilfull Physician presses by such cures, to purge and strengthen that part which is afflicted, as are only fit for that sort of disease, and doe best agree with the nature of that infirme part; which being abused to a disease of another nature, would prooue as hurtfull for the one, as helpfull for the other. Yea, not only will a skilfull and warie Physician bee carefull to vse no cure but that which is fit for that sort of disease, but he wil also consider all other circumstances, and make the remedies suitable thereunto; as the temperature of the clime where the Patient is, the constitution of the Planets,[7] the time of the Moone, the season of the yere, the age and complexion of the Patient, and the present state of his body, in strength or weaknesse. For one cure must not euer be vsed for the self-same disease, but according to the varying of any of the foresaid circumstances, that sort of remedie must be vsed which is fittest for the same. Whear by the contrarie in this case, such is the miraculous omnipotencie of our strong tasted *Tobacco*, as it cures all sorts of diseases (which neuer any drugge could do before) in all persons, and at all times. It cures all maner of distellations, either in the head or stomacke (if you beleeue their Axiomes) although in very deede it doe both corrupt the braine, and by causing ouer quicke disgestion, fill the stomacke full of crudities. It cures the Gowt in the feet, and (which is miraculous) in that very instant when the smoke thereof, as light, flies vp into the head, the vertue thereof, as heauie, runs downe to the little toe. It helpes all sorts of Agues. It makes a man sober that was drunke. It refreshes a weary man, and yet makes a man hungry. Being taken when they goe to bed, it makes one sleepe soundly, and yet being taken when a man is sleepie and drowsie, it will, as they say, awake his braine, and quicken his vnderstanding. As for curing of the Pockes, it serues for that vse but among the pockie Indian slaues. Here in *England* it is refined, and will not deigne to cure heere any other then cleanly and gentlemanly diseases. Omnipotent power of *Tobacco*! And if it could by the smoke thereof chace our deuils, as the smoke of *Tobias* fish did (which I am sure could smel no stronglier) it would serue for a precious Relicke, both for the supersti-

---

[7] This shows that so late as the 17th century the influence of the planets on the body was an article of firm belief, even amongst the learned. The following recipes may be of interest to the reader. They are taken from a manuscript volume which belonged to and was probably written by Sir John Floyer, physician to King Charles II., who practised at Lichfield, in the Cathedral library of which city the volume now is:—"An antidote to ye plague: take a cock chicken and pull off ye feathers from ye tayle till ye rump bee bare; you hold ye bare of ye same upon ye sore, and ye chicken will gape and labour for life, and in ye end will dye. Then take another and do ye like, and so another still as they dye, till one lives, for then ye venome is drawne out. The last chicken will live and ye patient will mend very speedily."

"Madness in a dog: 'Pega, Tega, Sega, Docemena Mega.' These words written, and ye paper rowl'd up and given to a dog, or anything that is mad, cure him."

tious Priests, and the insolent Puritanes, to cast out deuils withall. Admitting then, and not confessing that the vse thereof were healthfull for some sortes of diseases; should it be vsed for all sicknesses? should it be vsed by all men? should it be vsed at al times? yea should it be vsed by able, yong, strong, healthfull men? Medicine hath that vertue that it neuer leaueth a man in that state wherein it findeth him: it makes a sicke man whole, but a whole man sicke. And as Medicine helpes nature being taken at times of necessitie, so being euer and continually vsed, it doth but weaken, wearie, and weare nature. What speak I of Medicine? Nay let a man euery houre of the day, or as oft as many in this countrey vse to take *Tobacco*, let a man I say, but take as oft the best sorts of nourishments in meate and drinke that can bee deuised, hee shall with the continuall vse thereof weaken both his head and his stomacke: all his members shall become feeble, his spirits dull, and in the end, as a drowsie lazie belly-god, he shall euanish in a Lethargie.

And from this weaknesse it proceeds, that many in this kingdome haue had such a continuall vse of taking this vnsauerie smoke, as now they are not able to forbeare the same, no more than an olde drunkard can abide to be long sober, without falling into an vncurable weakenesse and euill constitution: for their continuall custome hath made to them, *habitum, alteram naturam*: so to those that from their birth haue bene continually nourished vpon poison and things venemous, wholesome meates are onely poisonable.

Thus hauing, as I truste, sufficiently answered the most principall arguments that are vsed in defence of this vile custome, it rests onely to informe you what sinnes and vanities you commit in the filthie abuse thereof. First are you not guiltie of sinnefull and shamefull lust? (for lust may bee as well in any of the senses as in feeling) that although you bee troubled with no disease, but in perfect health, yet can you neither be merry at an Ordinarie, nor lasciuious in the Stewes, if you lacke *Tobacco* to prouoke your appetite to any of those sorts of recreation, lusting after it as the children of Israel did in the wildernesse after Quailes? Secondly it is, as you vse or rather abuse it, a branche of the sinne of drunkennesse, which is the roote of all sinnes: for as the onely delight that drunkards take in wine is in the strength of the taste, and the force of the fume thereof that mounts vp to the braine: for no drunkards loue any weake, or sweete drinke: so are not those (I meane the strong heate and the fume), the onely qualities that make *Tobacco* so delectable to all the louers of it? And as no man likes strong headie drinke the first day (because *nemo repente fit turpissimus*), but by custome is piece and piece allured, while in the ende, a drunkard will haue as great a thirst with a draught as when hee hath need of it: So is not this the very case of all the great takers of *Tobacco*? which therefore they themselues do attribute to a bewitching qualitie in it. Thirdly, is it not the greatest sinne of all, that you the people of all sortes of this Kingdome, who are created and ordeined by God to bestowe both your persons and goods for the maintenance both of the honour and safetie of your King and Commonwealth, should disable yourselves in both? In your persons hauing by this continuall vile custome brought yourselues to this shameful imbecilitie, that you are not able to ride or walke the journey of a Jewes Sabboth, but you must haue a reekie cole brought you from the next poore house to kindle your *Tobacco* with? where as he cannot be thought able for any seruice in the warres, that cannot endure oftentimes the want of meate, drinke, and sleepe, much more then must hee endure the want of *Tobacco*. In the times of the many glorious and victorious battailes fought by this nation, there was no word of *Tobacco*. But now if it were time of warres, and that you were to make some sudden *Caualcado*[8] vpon your enemies, if any of you should seeke leisure to stay behinde his fellowe for taking of *Tobacco*, for my part I should neuer bee sorie for any euill chance that might befall him.[9] To take a custome in any thing that bee left againe, is most harmefull to the people of any land. *Mollicies* and delicacie were the wracke and ouerthrow, first of the Persian, and next of the Romane Empire. And this very custome of taking *Tobacco* (whereof our present purpose is), is euen at this day accounted so effeminate among the Indians themselues, as in the market they will offer no price for a slaue to be sold, whome they finde to be a great *Tobacco* taker.

    Now how you are by this custome disabled in your goods, let the gentry of this land beare witnesse, some of them bestowing three, some foure hundred pounds a yeere[10] vpon this precious stinke, which

---

[8] Or Camisado. A night attack on horseback, wherein the attacking party put their shirts on over their armour, in order to recognise each other in the darkness. Charles II. attempted a Camisado at Worcester, which did not succeed, owing to treachery.

[9] Our royal author would no doubt have been astonished to see English officers smoking on the field of battle, which I am told is now a common occurrence.

[10] It was not dreamt of in James's philosophy, that the price of tobacco might fall to 5s. 6d. and less a pound.

I am sure might be bestowed vpon many farre better vses. I read indeede of a knauish Courtier, who for abusing the fauour of the Emperour *Alexander Seuerus* his master by taking bribes to intercede, for sundry persons in his master's eare (for whom he neuer once opened his mouth) was iustly choked with smoke, with this doome, *Fumo pereat, qui fumum vendidit*: but of so many smoke-buyers, as are at this present in this kingdome, I neuer read nor heard.

And for the vanities committed in this filthie custome, is it not both great vanitie and vncleanenesse, that at the table, a place of respect, of cleanlinesse, of modestie, men should not be ashamed, to sit tossing of *Tobacco pipes*, and puffing of the smoke of *Tobacco* one to another, making the filthie smoke and stinke thereof, to exhale athwart the dishes, and infect the aire, when very often, men that abhorre it are at their repast? Surely Smoke becomes a kitchin far better then a Dining chamber, and yet it makes a kitchen also oftentimes in the inward parts of men, soiling and infecting them, with an vnctuous and oily kinde of Soote, as hath bene found in some great *Tobacco* takers, that after their death were opened. And not onely meate time, but no other time nor action is exempted from the publicke vse of this vnciuill tricke: so as if the wiues of *Diepe* list to contest with this nation for good maners their worst maners would in all reason be found at least not so dishonest (as ours are) in this point. The publike vse whereof, at all times, and in all places, hath now so farre preuailed, as diuers men very sound both in iudgement, and complexion, haue bene at last forced to take it also without desire, partly because they were ashamed to seeme singular (like the two Philosophers that were forced to duck themselues in that raine water, and so become fooles as well as the rest of the people) and partly, to be as one that was content to eate Garlicke (which he did not loue) that he might not be troubled with the smell of it, in the breath of his fellowes. And is it not a great vanitie, that a man cannot heartily welcome his friend now, but straight they must bee in hand with *Tobacco*? No it is become in place of a cure, a point of good fellowship, and he that will refuse to take a pipe of *Tobacco* among his fellowes, (though by his own election he would rather feele the sauour of a Sinke[11]) is accounted peeuish and no good company, euen as they doe with tippeling in the cold Easterne Countries. Yea the Mistresse cannot in a more manerly kinde, entertaine her seruant, then by giuing

---

[11] They still say in Scotland, "To feel a smell

him out of her faire hand a pipe of *Tobacco*. But herein is not onely a great vanitie, but a great contempt of God's good giftes, that the sweetenesse of mans breath, being a good gift of God, should be willfully corrupted by this stinking smoke, wherein I must confesse, it hath too strong a vertue: and so that which is an ornament of nature, and can neither by any artifice be at the first acquired, nor once lost, be recouered againe, shall be filthily corrupted with an incurable stinke, which vile qualitie is as directly contrary to that wrong opinion which is holden of the wholesomnesse thereof, as the venime of putrifaction is contrary to the vertue Preseruatiue.

Moreouer, which is a great iniquitie, and against all humanitie, the husband shall not bee ashamed, to reduce thereby his delicate, wholesome, and cleane complexioned wife, to that extremetie, that either shee must also corrupt her sweete breath therewith, or else resolue to liue in a perpetuall stinking torment.

Haue you not reason then to bee ashamed, and to forbeare this filthie noueltie, so basely grounded, so foolishly receiued and so grossely mistaken in the right vse thereof? In your abuse thereof sinning against God, harming yourselues both in persons and goods, and taking also thereby the markes and notes of vanitie vpon you: by the custome thereof making your selues to be wondered at by all forraine ciuil Nations, and by all strangers that come among you, to be scorned and contemned. A custome lothsome to the eye, hatefull to the Nose, harmefull to the braine, dangerous to the Lungs, and in the blacke stinking fume thereof, nearest resembling the horrible Stigian smoke of the pit that is bottomelesse.

Daemonologie

by
King James I of England

Preface

    The fearefull aboundinge at this time in this countrie, of these detestable slaues of the Deuill, the Witches or enchaunters, hath moved me (beloued reader) to dispatch in post, this following treatise of mine, not in any wise (as I protest) to serue for a shew of my learning & ingine, but onely (mooued of conscience) to preasse thereby, so farre as I can, to resolue the doubting harts of many; both that such assaultes of Sathan are most certainly practized, & that the instrumentes thereof, merits most severly to be punished: against the damnable opinions of two principally in our age, wherof the one called SCOT an Englishman, is not ashamed in publike print to deny, that ther can be such a thing as Witch-craft: and so mainteines the old error of the Sadducees, in denying of spirits. The other called VVIERVS, a German Phisition, sets out a publick apologie for al these craftes-folkes, whereby, procuring for their impunitie, he plainely bewrayes himselfe to haue bene one of that profession. And for to make this treatise the more pleasaunt and facill, I haue put it in forme of a Dialogue, which I haue diuided into three bookes: The first speaking of Magie in general, and Necromancie in special. The second of Sorcerie and Witch-craft: and the thirde, conteines a discourse of all these kindes of spirits, & Spectres that appeares & trobles persones: together with a conclusion of the whol work. My intention in this labour, is only to proue two things, as I haue alreadie said: the one, that such diuelish artes haue bene and are. The other, what exact trial and seuere punishment they merite: & therefore reason I, what kinde of things are possible to be performed in these arts, & by what naturall causes they may be, not that I touch every particular thing of the Deuils power, for that were infinite: but onelie, to speak scholasticklie, (since this can not bee spoken in our language) I reason vpon *genus* leauing species, *and differentia* to be comprehended therein. As for example, speaking of the power of Magiciens, in the first book & sixt Chapter: I say, that they can suddenly cause be brought vnto them, all kindes of daintie disshes, by their familiar spirit: Since as a thiefe he delightes to steale, and as a spirite, he can subtillie & suddenlie inough transport the same. Now vnder this *genus* may be comprehended al particulars, depending thereupon; Such as the bringing Wine out of a Wall, (as we haue heard oft to haue bene practised] and such others; which particulars, are sufficientlie proved by the reasons of the general. And such like in the second booke of Witch-craft in speciall, and fift Chap. I say and proue by diuerse arguments, that Witches can, by the power of their Master, cure or cast on disseases: Now by these same reasones, that proues their power by the Deuil of disseases in generally is aswell proued their power in speciall: as of weakening the nature of some men, to make them vnable for women: and making it to abound in others, more then the ordinary course of nature would permit. And such like in all other particular sicknesses; But one thing I will pray thee to obserue in all these places, where I reason vpon the deuils power, which is the different ends & scopes, that God as the first cause, and the Devill as his instrument and second cause shootes at in all these actiones of the Deuil, (as Gods hang-man:) For where the deuilles intention in them is euer to perish, either the soule or the body, or both of them, that he is so permitted to deale with: God by the contrarie, drawes euer out of that euill glorie to himselfe, either by the wracke of the wicked in his justice, or by the tryall of the patient, and amendment of the faithfull, being wakened vp with that rod of correction. Hauing thus declared vnto thee then, my full intention in this Treatise, thou wilt easelie excuse, I doubt not, aswel my pretermitting, to declare the whole particular rites and secretes of these vnlawfull artes: as also their infinite and wounderfull practises, as being neither of them pertinent to my purpose: the reason whereof, is giuen in the hinder ende of the first Chapter of the thirde booke: and who likes to be curious in these thinges, he may reade, if he will here of their practises, BODINVS Dæmonomanie, collected with greater diligence, then written with judgement, together with their confessions, that haue bene at this time apprehened. If he would know what hath bene the opinion of the Auncientes, concerning their power: he shall see it wel described by HYPERIVS, & HEMMINGIVS, two late Germaine writers: Besides innumerable other neoterick Theologues, that writes largelie vpon that subject: And if he woulde knowe what are the particuler rites, & curiosities of these black arts (which is both vnnecessarie and perilous,) he will finde it in the fourth book of CORNELIVS Agrippa, and in VVIERVS, whomof I spak. And so wishing my pains in this Treatise (beloued Reader} to be

effectual, in arming al them that reades the same, against these aboue mentioned erroures, and recommending my good will to thy friendly acceptation, I bid thee hartely fare-well.
    IAMES R^x.

First Booke

ARGVMENT.

The exord of the whole. The description of Magie in speciall.

Chapter 1

ARGVMENT.

Proven by the Scripture, that these vnlawfull artes in genere, *haue bene and may be put in practise.*

PHILOMATHES and EPISTEMON reason the matter.

PHILOMATHES.

I am surely verie glad to haue mette with you this daye, for I am of opinion, that ye can better resolue me of some thing, wherof I stand in great doubt, nor anie other whom-with I could haue mette.

EPI. In what I can, that ye like to speir at me, I will willinglie and freelie tell my opinion, and if I proue it not sufficiently, I am heartely content that a better reason carie it away then.

PHI. What thinke yee of these strange newes, which now onelie furnishes purpose to al men at their meeting: I meane of these Witches?

EPI. Surelie they are wonderfull: And I think so cleare and plaine confessions in that purpose, haue neuer fallen out in anie age or cuntrey.

PHI. No question if they be true, but thereof the Doctours doubtes.

EPI. What part of it doubt ye of?

PHI. Even of all, for ought I can yet perceaue: and namelie, that there is such a thing as Witch-craft or Witches, and I would pray you to resolue me thereof if ye may: for I haue reasoned with sundrie in that matter, and yet could never be satisfied therein.

EPI. I shall with good will doe the best I can: But I thinke it the difficiller, since ye denie the thing it selfe in generall: for as it is said in the logick schools, *Contra negantem principia non est disputandum.* Alwaies for that part, that witchcraft, and Witches haue bene, and are, the former part is clearelie proved by the Scriptures, and the last by dailie experience and confessions.

PHI. I know yee will alleadge me *Saules Pythonisse*: but that as appeares will not make much for you.

EPI. Not onlie that place, but divers others: But I marvel why that should not make much for me?

PHI. The reasons are these, first yee may consider, that *Saul* being troubled in spirit,
1. *Sam.* 28.
and having fasted long before, as the text testifieth, and being come to a woman that was bruted to have such knowledge, and that to inquire so important news, he having so guiltie a conscience for his hainous offences, and specially, for that same vnlawful curiositie, and horrible defection: and then the woman crying out vpon the suddaine in great admiration, for the vncouth sicht that she alledged to haue sene, discovering him to be the King, thogh disguysed, & denied by him before: it was no wounder I say, that his senses being thus distracted, he could not perceaue hir faining of hir voice, hee being himselfe in an other chalmer, and seeing nothing. Next what could be, or was raised? The spirit of *Samuel*? Prophane and against all Theologie: the Diuell in his likenes? as vnappeirant, that either God would permit him to come in the shape of his Saintes (for then could neuer the Prophets in those daies haue bene sure, what Spirit spake to them in their visiones) or then that he could fore-tell what was to come there after; for Prophecie proceedeth onelie of GOD: and the Devill hath no knowledge of things to come.

EPI. Yet if yee will marke the wordes of the text, ye will finde clearely, that *Saul* saw that apparition: for giving you that *Saul* was in an other Chalmer, at the making of the circles & conjurationes, needeful for that purpose (as none of that craft will permit any vthers to behold at that time) yet it is evident by the text, that how sone that once that vnclean spirit was fully risen, shee called in vpon *Saul*. For it is saide in the text, that *Saule knew him to be Samuel*, which coulde not haue bene, by the hearing tell onely of an olde man with an mantil, since there was many mo old men dead in *Israel* nor *Samuel*: And the common weid of that whole Cuntrey was mantils. As to the next, that it was not the spirit of *Samuel*, I grant: In the proving whereof ye neede not to insist, since all Christians of whatso-ever Religion agrees vpon that: and none but either mere ignorants, or Necromanciers or Witches doubtes thereof. And that the Diuel is permitted at som-times to put himself in the liknes of the Saintes, it is plaine in the Scriptures,

DAEMONOLOGIE

where it is said, that *Sathan can trans-forme himselfe into an Angell of light.*
    2. *Cor.* 11.14.

Neither could that bring any inconvenient with the visiones of the Prophets, since it is most certaine, that God will not permit him so to deceiue his own: but only such, as first wilfully deceiues themselues, by running vnto him, whome God then suffers to fall in their owne snares, and justlie permittes them to be illuded with great efficacy of deceit, because they would not beleeue the trueth (as *Paul* sayth). And as to the diuelles foretelling of things to come, it is true that he knowes not all things future, but yet that he knowes parte, the Tragicall event of this historie declares it, (which the wit of woman could never haue fore-spoken) not that he hath any prescience, which is only proper to God: or yet knows anie thing by loking vpon God, as in a mirrour (as the good Angels doe) he being for euer debarred from the fauorable presence & countenance of his creator, but only by one of these two meanes, either as being worldlie wise, and taught by an continuall experience, ever since the creation, judges by likelie-hood of thinges to come, according to the like that hath passed before, and the naturall causes, in respect of the vicissitude of all thinges worldly: Or else by Gods employing of him in a turne, and so foreseene thereof: as appeares to haue bin in this, whereof we finde the verie like in *Micheas* propheticque discourse to King *Achab.*
    1. *King.* 22.

But to prooue this my first proposition, that there can be such a thing as witch-craft, & witches, there are manie mo places in the Scriptures then this (as I said before). As first in the law of God, it is plainely prohibited:
    Exod. 22.

But certaine it is, that the Law of God speakes nothing in vaine, nether doth it lay curses, or injoyne punishmentes vpon shaddowes, condemning that to be il, which is not in essence or being as we call it. Secondlie it is plaine, where wicked *Pharaohs* wise-men imitated ane number of *Moses* miracles,
    Exod. 7 & 8.

to harden the tyrants heart there by. Thirdly, said not *Samuell* to *Saull*,
    1. *Sam.* 15.

that *disobedience is as the sinne of Witch-craft*? To compare to a thing that were not, it were too too absurd. Fourthlie, was not *Simon Magus*, a man of that craft?

Acts. 8.
And fiftlie, what was she that had the spirit of *Python*?
Acts 16.
beside innumerable other places that were irkesom to recite.

Chapter 2

ARGVMENT.
What kynde of sin the practizers of these vnlawfull artes committes. The division of these artes. And what are the meanes that allures any to practize them.
PHILOMATHES.
Bvt I thinke it very strange, that God should permit anie man-kynde (since they beare his owne Image) to fall in so grosse and filthie a defection.
EPI. Although man in his Creation was
Gen. 1.
made to the Image of the Creator, yet through his fall having once lost it, it is but restored againe in a part by grace onelie to the elect: So all the rest falling away from God, are given over in the handes of the Devill that enemie, to beare his Image: and being once so given over, the greatest and the grossest impietie, is the pleasantest, and most delytefull vnto them.
PHI. But may it not suffice him to haue indirectly the rule, and procure the perdition of so manie soules by alluring them to vices, and to the following of their own appetites, suppose he abuse not so many simple soules, in making them directlie acknowledge him for their maister.
EPI. No surelie, for hee vses everie man, whom of he hath the rule, according to their complexion and knowledge: And so whome he findes most simple, he plaineliest discovers himselfe vnto them. For hee beeing the enemie of mans Salvation, vses al the meanes he can to entrappe them so farre in his snares, as it may be vnable to them thereafter (suppose they would) to rid themselues out of the same.
PHI. Then this sinne is a sinne against the holie Ghost.
EPI. It is in some, but not in all.

PHI. How that? Are not all these that runnes directlie to the Devill in one Categorie.

EPI. God forbid, for the sin against the holie Ghost hath two branches: The one a falling backe from the whole service of GOD, and a refusall of all his preceptes. The other is the doing of the first with knowledge, knowing that they doe wrong against their own conscience, and the testimonie of

Heb. 6. 10.

the holie Spirit, having once had a tast of the sweetnes of Gods mercies. Now in the first of these two, all sortes of Necromancers, Enchanters or Witches, ar comprehended: but in the last, none but such as erres with this knowledge that I haue spoken of.

PHI. Then it appeares that there are more sortes nor one, that are directlie professors of his service: and if so be, I pray you tell me how manie, and what are they?

EPI. There are principallie two sortes, wherevnto all the partes of that vnhappie arte are redacted; whereof the one is called *Magie* or *Necromancie*, the other *Sorcerie* or *Witch-craft*.

PHI. What I pray you? and how manie are the meanes, whereby the Devill allures persones in anie of these snares? EPI. Even by these three passiones that are within our selues: Curiositie in great ingines: thrist of revenge, for some tortes deeply apprehended: or greedie appetite of geare, caused through great pouerty. As to the first of these, Curiosity, it is onelie the inticement of *Magiciens*, or *Necromanciers*: and the other two are the allureres of the *Sorcerers*, or *Witches*, for that olde and craftie Serpent, being a spirite, hee easilie spyes our affections, and so conformes himselfe thereto, to deceaue vs to our wracke.

Chapter 3

ARGVMENT.

The significations and Etymologies of the words of Magie *and* Necromancie. *The difference betuixt* Necromancie *and* Witch-craft: *What are the entressis, and beginninges, that brings anie to the knowledge thereof.*

PHILOMATHES.

I Would gladlie first heare, what thing is it that ye call *Magie* or *Necromancie*.

EPI. This worde *Magie* in the *Persian* toung, importes as muche as to be ane contemplator or Interpretour of Divine and heavenlie sciences: which being first vsed amongs the *Chaldees*, through their ignorance of the true divinitie, was esteemed and reputed amongst them, as a principall vertue: And therefore, was named vnjustlie with an honorable stile, which name the *Greekes* imitated, generally importing all these kindes of vnlawfull artes. And this word *Necromancie* is a Greek word, compounded of Νεκρων & μαντεια, which is to say, the Prophecie by the dead. This last name is given, to this black & vnlawfull science by the figure *Synedoche*, because it is a principal part of that art, to serue them selues with dead carcages in their diuinations.

Phi. What difference is there betwixt this arte, and Witch-craft.

EPI. Surelie, the difference vulgare put betwixt them, is verrie merrie, and in a maner true; for they say, that the Witches ar servantes onelie, and slaues to the Devil; but the Necromanciers are his maisters and commanders.

PHI. How can that be true, yt any men being specially adicted to his service, can be his commanders?

EPI. Yea, they may be: but it is onelie *secundum quid*: For it is not by anie power that they can haue over him, but *ex pacto* allanerlie:

whereby he oblices himself in some trifles to them, that he may on the other part obteine the fruition of their body & soule, which is the onlie thing he huntes for.

PHI. An verie in-æquitable contract forsooth: But I pray you discourse vnto mee, what is the effect and secreets of that arte?

EPI. That is over large an fielde ye giue mee: yet I shall doe good-will, the most summarlie that I can, to runne through the principal points thereof. As there are two sorts of folkes, that may be entysed to this arte, to wit, learned or vnlearned: so is there two meanes, which are the first steerers vp & feeders of their curiositie, thereby to make them to giue themselves over to the same: Which two meanes, I call the Divels schoole, and his rudimentes. The learned haue their curiositie wakened vppe; and fedde by that which I call his schoole: this is the *Astrologie* judiciar. For divers men having attained to a great perfection in learning, & yet remaining overbare (alas) of the spirit of regeneration and frutes thereof: finding all naturall thinges common, aswell to the stupide pedants as vnto them, they assaie to vendicate vnto them a greater name, by not onlie knowing the course of things heavenlie, but likewise to cling to the knowledge of things to come thereby. Which, at the first face appearing lawfull vnto them, in respect the ground therof seemeth to proceed of naturall causes onelie: they are so allured thereby, that finding their practize to prooue true in sundry things, they studie to know the cause thereof: and so mounting from degree to degree, vpon the slipperie and vncertaine scale of curiositie; they are at last entised, that where lawfull artes or sciences failes, to satisfie their restles mindes, even to seeke to that black and vnlawfull science of *Magie*. Where, finding at the first, that such diuers formes of circles & conjurations rightlie joyned thereunto, will raise such divers formes of spirites, to resolue them of their doubts: and attributing the doing thereof, to the power inseparablie tyed, or inherent in the circles: and manie words of God, confusedlie wrapped in; they blindlie glorie of themselves, as if they had by their quicknes of ingine, made a conquest of *Plutoes* dominion, and were become Emperours over the *Stygian* habitacles. Where, in the meane time (miserable wretches) they are become in verie deede, bond-slaues to their mortall enemie: and their knowledge, for all that they presume thereof, is nothing increased, except in knowing evill, and the horrors of Hell for punishment thereof, as *Adams* was by the eating of the forbidden tree.

Gen. 3.

Chapter 4

ARGVMENT.
The Description of the Rudiments and Schoole, which are the entresses to the arte of Magie: *And in speciall the differences betwixt* Astronomie *and* Astrologie: *Diuision of* Astrologie *in diuers partes.*

PHILOMATHES.
Bvt I pray you likewise forget not to tell what are the Deuilles rudimentes.

EPI. His rudimentes, I call first in generall, all that which is called vulgarly the vertue of worde, herbe, & stone: which is vsed by vnlawful charmes, without naturall causes. As likewise all kinde of practicques, freites, or other like extraordinarie actiones, which cannot abide the true toutche of naturall reason.

PHI. I would haue you to make that playner, by some particular examples; for your proposition is verie generall.

EPI. I meane either by such kinde of Charmes as commonlie dafte wiues vses, for healing of forspoken goodes, for preseruing them from euill eyes, by knitting roun-trees, or sundriest kinde of herbes, to the haire or tailes of the goodes: By curing the Worme, by stemming of blood, by healing of Horse-crookes, by turning of the riddle, or doing of such like innumerable things by wordes, without applying anie thing, meete to the part offended, as Mediciners doe; Or else by staying maried folkes, to haue naturallie adoe with other, (by knitting so manie knottes vpon a poynt at the time of their mariage). And such-like things, which men vses to practise in their merrinesse: For fra vnlearned men (being naturallie curious, and lacking the true knowledge of God) findes these practises to prooue true, as sundrie of them will doe, by the power of the Devill for deceauing men, and not by anie inherent vertue in these vaine wordes and freites; & being desirous to winne a reputation to themselues in such-like turnes, they either (if they be of the shamefaster sorte) seeke to bee learned by some that are experimented in that Arte, (not knowing it to be euill at the first) or else being of the grosser sorte, runnes directlie to the Deuill for ambition or desire of gaine, and plainelie contractes with him thereupon.

PHI. But me thinkes these meanes which yee call the Schoole and rudimentes of the Deuill, are thinges lawfull, and haue bene approoued for such in all times and ages: As in special, this science of *Astrologie*, which is one of the speciall members of the *Mathematicques*.

EPI. There are two thinges which the learned haue obserued from the beginning, in the science of the Heauenlie Creatures, the Planets, Starres, and such like: The one is their course and ordinary motiones, which for that cause is called *Astronomia*: Which word is a compound of νομος & αστερων that is to say, the law of the Starres: And this arte indeed is one of the members of the *Mathematicques*, & not onelie lawful, but most necessarie and commendable. The other is called *Astrologia*, being compounded of αστερων & λογος which is to say, the word, and preaching of the starres: Which is deuided in two partes: The first by knowing thereby the powers of simples, and sickenesses, the course of the seasons and the weather, being ruled by their influence; which part depending vpon the former, although it be not of it selfe a parte of *Mathematicques*: yet it is not vnlawful, being moderatlie vsed, suppose not so necessarie and commendable as the former. The second part is to truste so much to their influences, as thereby to fore-tell what common-weales shall florish or decay: what persones shall be fortunate or vnfortunate: what side shall winne in anie battell: What man shall obteine victorie at singular combate: What way, and of what age shall men die: What horse shall winne at matche-running; and diuerse such like incredible things, wherein *Cardanus*, *Cornelius Agrippa*, and diuerse others haue more curiouslie then profitably written at large. Of this roote last spoken of, springs innumerable branches; such as the knowledge by the natiuities; the *Cheiromancie*, *Geomantie*, *Hydromantie*, *Arithmantie*, *Physiognomie*: & a thousand others: which were much practised, & holden in great reuerence by the *Gentles* of olde. And this last part of *Astrologie* whereof I haue spoken, which is the root of their branches, was called by them *pars fortunæ*. This parte now is vtterlie vnlawful to be trusted in, or practized amongst christians, as leaning to no ground of natural reason: & it is this part which I called before the deuils schole.

PHI. But yet manie of the learned are of the contrarie opinion.

DAEMONOLOGIE

 EPI. I grant, yet I could giue my reasons to fortifie & maintaine my opinion, if to enter into this disputation it wold not draw me quite off the ground of our discours; besides the mis-spending of the whole daie thereupon: One word onely I will answet to them, & that in the Scriptures (which must be an infallible ground to all true Christians) That in the Prophet *Ieremie*
 Ierem. 10.
 it is plainelie forbidden, to beleeue or hearken vnto them that Prophecies & fore-speakes by the course of the Planets & Starres.

Chapter 5

ARGVMENT.

How farre the vsing of Charmes is lawfull or vnlawfull: The description of the formes of Circkles and Coniurationes. And what causeth the Magicianes *themselues to wearie thereof.*

PHILOMATHES.

Wel, Ye haue said far inough in that argument. But how prooue ye now that these charmes or vnnaturall practicques are vnlawfull: For so, many honest & merrie men & women haue publicklie practized some of them, that I thinke if ye would accuse them al of Witch-craft, ye would affirme more nor ye will be beleeued in.

EPI. I see if you had taken good tent (to the nature of that word, whereby I named it,) ye would not haue bene in this doubt, nor mistaken me, so farre as ye haue done: For although, as none can be schollers in a schole, & not be subject to the master thereof: so none can studie and put in practize (for studie the alone, and knowledge, is more perilous nor offensiue; and it is the practise only that makes the greatnes of the offence.) the cirkles and art of *Magie*, without committing an horrible defection from God: And yet as they that reades and learnes their rudiments, are not the more subject to anie schoole-master, if it please not their parentes to put them to the schoole thereafter; So they who ignorantly proues these practicques, which I cal the deuilles rudiments, vnknowing them to be baites, casten out by him, for trapping such as God will permit to fall in his hands: This kinde of folkes I saie, no doubt, ar to be judged the best of, in respect they vse no invocation nor help of him (by their knowledge at least) in these turnes, and so haue neuer entred themselues in Sathans seruice; Yet to speake truely for my owne part (I speake but for my selfe) I desire not to make so neere riding: For in my opinion our enemie is ouer craftie, and we ouer weake (except the greater

DAEMONOLOGIE

grace of God) to assay such hazards, wherein he preases to trap vs.

PHI. Ye haue reason forsooth; for as the common Prouerbe saith: They that suppe keile with the Deuill, haue neede of long spoones. But now I praie you goe forwarde in the describing of this arte of *Magie*.

EPI. Fra they bee come once vnto this perfection in euill, in hauing any knowledge (whether learned or vnlearned) of this black art: they then beginne to be wearie of the raising of their Maister, by conjured circkles; being both so difficile and perilous, and so commeth plainelie to a contract with him, wherein is speciallie conteined formes and effectes.

PHI. But I praye you or euer you goe further, discourse me some-what of their circkles and conjurationes; And what should be the cause of their wearying thereof: For it should seeme that that forme should be lesse fearefull yet, than the direct haunting and societie, with that foule and vncleane Spirite.

EPI. I thinke ye take me to be a Witch my selfe, or at the least would faine sweare your selfe prentise to that craft: Alwaies as I may, I shall shortlie satisfie you, in that kinde of conjurations, which are conteined in such bookes, which I call the Deuilles Schoole: There are foure principall partes; the persons of the conjurers; the action of the conjuration; the wordes and rites vsed to that effect; and the Spirites that are conjured. Ye must first remember to laye the ground, that I tould you before: which is, that it is no power inherent in the circles, or in the holines of the names of God blasphemouslie vsed: nor in whatsoeuer rites or ceremonies at that time vsed, that either can raise any infernall spirit, or yet limitat him perforce within or without these circles. For it is he onelie, the father of all lyes, who hauing first of all prescribed that forme of doing, feining himselfe to be commanded & restreined thereby, wil be loath to passe the boundes of these injunctiones; aswell thereby to make them glory in the impiring ouer him (as I saide before:) As likewise to make himselfe so to be trusted in these little thinges, that he may haue the better commoditie thereafter, to deceiue them in the end with a tricke once for all; I meane the euerlasting perdition of their soul & body. Then laying this ground, as I haue said, these conjurationes must haue few or mo in number of the persones conjurers (alwaies passing the singuler number) according to the qualitie of the circle, and forme of apparition. Two principall thinges cannot well in that errand be wanted: holie-water (whereby the Deuill mockes the *Papistes*) and some present of a liuing thing vnto him. There ar likewise certaine seasons, dayes and houres, that they obserue in this purpose: These things being all readie, and prepared, circles are made triangular, quadrangular, round, double or single, according to the forme of apparition that they craue. But to speake of the diuerse formes of the circles, of the innumerable characters and crosses that are within and without, and out-through the same, of the diuers formes of apparitiones, that that craftie spirit illudes them with, and or all such particulars in that action, I remit it to ouer-manie that haue busied their heades in describing of the same; as being but curious, and altogether vnprofitable. And this farre onelie I touch, that when the conjured Spirit appeares, which will not be while after manie circumstances, long praiers, and much muttring and murmuring of the conjurers; like a *Papist* priest, dispatching a hunting *Masse*: how sone I say, he appeares, if they haue missed one iote of all their rites; or if any of their feete once slyd ouer the circle through terror of his feareful apparition, he payes himselfe at that time in his owne hande, of that due debt which they ought him; and other-wise would haue delayed longer to haue payed him: I meane hee carries them with him bodie and soule. If this be not now a just cause to make them wearie of these formes of conjuration, I leaue it to you to judge vpon; considering the long-somenesse of the labour, the precise keeping of dayes and houres (as I haue said), the terriblenesse of apparition, and the present perrell that they stande in, in missing the least circumstance or freite, that they ought to obserue: And on the other parte, the Deuil is glad to mooue them to a plaine and square dealing with him as I said before.

Chapter 6

ARGVMENT.

The Deuilles contract with the Magicians: The diuision thereof in two partes: What is the difference betwixt Gods miracles and the Deuils.

PHILOMATHES.

Indeede there is cause inough, but rather to leaue him at all, then to runne more plainlie to him, if they were wise he delt with. But goe forwards now I pray you to these turnes, fra they become once deacons in this craft.

EPI. From time that they once plainelie begin to contract with him: The effect of their contract consistes in two thinges; in formes and effectes, as I began to tell alreadie, were it not yee interrupted me (for although the contract be mutuall; I speake first of that part, wherein the Deuill oblishes himselfe to them) by formes, I meane in what shape or fashion he shall come vnto them, when they call vpon him. And by effectes, I vnderstand, in what special sorts of seruices he bindes himselfe to be subject vnto them. The qualitie of these formes and effectes, is lesse or greater, according to the skil and art of the *Magician*. For as to the formes, to some of the baser sorte of them he oblishes him selfe to appeare at their calling vpon him, by such a proper name which he shewes vnto them, either in likenes of a dog, a Catte, an Ape, or such-like other beast; or else to answere by a voyce onlie. The effects are to answere to such demands, as concernes curing of disseases, their own particular menagery: or such other base things as they require of him. But to the most curious sorte, in the formes he will oblish himselfe, to enter in a dead bodie, and there out of to giue such answers, of the euent of battels, of maters concerning the estate of commonwelths, and such like other great questions: yea, to some he will be a continuall attender, in forme of a Page: He will permit himselfe to be conjured, for the space of so many yeres, ether in a tablet or a ring, or such like thing, which they may easely carrie about with them: He giues them power to sel such wares to others, whereof some will bee dearer, and some better cheape; according to the lying or true speaking of the Spirit that is conjured therein. Not but that in verie deede, all Devils must be lyars; but so they abuse the simplicitie of these wretches, that becomes their schollers, that they make them beleeue, that at the fall of *Lucifer*, some Spirites fell in the aire, some in the fire, some in the water, some in the lande: In which Elementes they still remaine. Whereupon they build, that such as fell in the fire, or in the aire, are truer then they, who fell in the water or in the land, which is al but meare trattles, & forged by the author of al deceit. For they fel not be weight, as a solide substance, to stick in any one parte: But the principall part of their fal, consisting in qualitie, by the falling from the grace of God wherein they were created, they continued still thereafter, and shal do while the latter daie, in wandring through the worlde, as Gods hang-men, to execute such turnes as he employes them in. And when anie of them are not occupyed in that, returne they must to their prison in hel (as it is plaine in the miracle that CHRIST wrought at *Gennezareth*)

Mat. 8.

therein at the latter daie to be all enclosed for euer: and as they deceiue their schollers in this, so do they, in imprinting in them the opinion that there are so manie Princes, Dukes, and Kinges amongst them, euerie one commanding fewer or mo Legions, and impyring in diuers artes, and quarters of the earth. For though that I will not denie that there be a forme of ordour amongst the Angels in Heauen, and consequentlie, was amongst them before their fall; yet, either that they bruike the same sensine; or that God will permit vs to know by damned Deuils, such heauenlie mysteries of his, which he would not reueale to vs neither by Scripture nor Prophets, I thinke no Christiane will once thinke it. But by the contrarie of all such mysteries, as he hath closed vp with his seale of secrecie; it becommeth vs to be contented with an humble ignorance, they being thinges not necessarie for our saluation. But to returne to the purpose, as these formes, wherein Sathan oblishes himselfe to the greatest of the *Magicians*, are wounderfull curious; so are the effectes correspondent vnto the same: For he will oblish himselfe to teach them artes and sciences, which he may easelie doe, being so learned a knaue as he is: To carrie them newes from anie parte of the worlde, which the agilitie of a Spirite may easelie performe: to reueale to them the secretes of anie persons, so being they bee once spoken, for the thought none knowes but GOD;

except so far as yee may ghesse by their countenance, as one who is doubtleslie learned, inough in the *Physiognomie*: Yea, he will make his schollers to creepe in credite with Princes, by fore-telling them manie greate thinges; parte true, parte false: For if all were false, he would tyne credite at all handes; but alwaies doubtsome, as his Oracles were. And he will also make them to please Princes, by faire banquets and daintie dishes, carryed in short space fra the farthest part of the worlde. For no man doubts but he is a thiefe, and his agilitie (as I spake before) makes him to come suche speede. Such-like, he will guard his schollers with faire armies of horse-men and foote-men in appearance, castles and fortes: Which all are but impressiones in the aire, easelie gathered by a spirite, drawing so neare to that substance himselfe: As in like maner he will learne them manie juglarie trickes at Gardes, dice, & such like, to deceiue mennes senses thereby: and such innumerable false practicques; which are prouen by ouer-manie in this age: As they who ar acquainted with that *Italian* called SCOTO yet liuing, can reporte. And yet are all these thinges but deluding of the senses, and no waies true in substance, as were the false miracles wrought by King *Pharaoes* Magicians, for counterfeiting *Moyses*: For that is the difference betuixt Gods myracles and the Deuils, God is a creator, what he makes appeare in miracle, it is so in effect. As *Moyses* rod being casten downe, was no doubt turned in a natural Serpent: where as the Deuill (as Gods Ape) counterfetting that by his *Magicians*, maid their wandes to appeare so, onelie to mennes outward senses: as kythed in effect by their being deuoured by the other. For it is no wonder, that the Deuill may delude our senses, since we see by common proofe, that simple juglars will make an hundreth thinges seeme both to our eies and eares otherwaies then they are. Now as to the *Magicians* parte of the contract, it is in a word that thing, which I said before, the Deuill hunts for in all men.

PHI. Surelie ye haue said much to me in this arte, if all that ye haue said be as true as wounderfull.

EPI. For the trueth in these actiones, it will be easelie confirmed, to anie that pleases to take paine vpon the reading of diuerse authenticque histories, and the inquiring of daily experiences. And as for the trueth of their possibilitie, that they may be, and in what maner, I trust I haue alleaged nothing whereunto I haue not joyned such probable reasons, as I leaue to your discretion, to waie and consider: One word onlie I omitted; concerning the forme of making of this contract, which is either written with the *Magicians* owne bloud: or else being agreed vpon (in termes his schole-master) touches him in some parte, though peraduenture no marke remaine: as it doth with all Witches.

Chapter 7

ARGVMENT.

The reason why the art of Magie *is unlawfull. What punishment they merite: And who may be accounted guiltie of that crime.*

PHILOMATHES.

Svrelie Ye haue made this arte to appeare verie monstruous & detestable. But what I pray you shall be said to such as mainteines this art to be lawfull, for as euill as you haue made it?

EPI. I say, they sauour of the panne them selues, or at least little better, And yet I would be glad to heare their reasons.

PHI. There are two principallie, that euer I heard vsed; beside that which is founded vpon the common Prouerb (that the *Necromancers* commands the Deuill, which ye haue already refuted). The one is grounded vpon a receiued custome: The other vpon an authoritie, which some thinkes infallible. Vpon custome, we see that diuerse Christian Princes and Magistrates seuere punishers of Witches, will not onelie ouer-see *Magicians* to liue within their dominions; but euen some-times delight to see them prooue some of their practicques. The other reason is, that *Moyses* being brought vp (as it is expreslie said in the Scriptures) *in all the sciences of the Ægyptians*; whereof no doubt, this was one of the principalles. And he notwithstanding of this arte, pleasing God, as he did, consequentlie that art professed by so godlie a man, coulde not be vnlawfull. EPI. As to the first of your reasones, grounded vpon custome: I saie, an euill custome can neuer be accepted for a good law, for the ouer great ignorance of the worde in some Princes and Magistrates, and the contempt thereof in others, moues them to sinne heauelie against their office in that poynt. As to the other reasone, which seemes to be of greater weight, if it were formed in a Syllogisme; it behooued to be in manie termes, and full of fallacies (to speake in termes of *Logicque*) for first, that that generall proposition; affirming *Moyses* to be taught *in all the sciences of the Ægyptians*, should conclude that he was taught in *Magie*, I see no necessity. For we must vnderstand that the spirit of God there, speaking of sciences, vnderstandes them that are lawfull; for except they be lawfull, they are but *abusiuè* called sciences, & are but ignorances indeede: *Nam homo pictus, non est homo*. Secondlie, giuing that he had bene taught in it, there is great difference, betwixt knowledge and practising of a thing (as I said before). For God knoweth all thinges, being alwaies good, and of our sinne & our infirmitie proceedeth our ignorance. Thirdlie, giuing that he had both studied and practised the same (which is more nor monstruous to be beleeued by any Christian) yet we know well inough, that before that euer the spirite of God began to call *Moyses*, he was fled out of *Ægypt*, being fourtie yeares of age, for the slaughter of an *Ægyptian*, and in his good-father *Iethroes* lande, first called at the firie bushe, hauing remained there other fourtie yeares in exile: so that suppose he had beene the wickeddest man in the worlde before, he then became a changed and regenerat man, and very litle of olde *Moyses* remained in him. *Abraham* was an Idolater in *Vr* of *ChaldÃ|æa*, before he was called: And *Paule* being called *Saule*, was a most sharp persecutor of the Saintes of God, while that name was changed.

PHI. What punishment then thinke ye merites these *Magicians* and *Necromancers*?

EPI. The like no doubt, that *Sorcerers* and *Witches* merites; and rather so much greater, as their error proceedes of the greater knowledge, and so drawes nerer to the sin against the holy Ghost. And as I saye of them, so saye I the like of all such as consults, enquires, entertaines, & ouersees them, which is seene by the miserable endes of many that askes councell of them: For the Deuill hath neuer better tydings to tell to any, then he tolde to *Saule*: neither is it lawfull to vse so vnlawfull instrumentes, were it neuer for so good a purpose: for that axiome in Theologie is most certaine and infallible:

Ast 3.
Nunquam faciendum est malum vt bonum inde eueniat.

Seconde Booke

ARGVMENT.

The description of Sorcerie and Witchcraft in speciall.

Chapter 1

ARGVMENT.

Proued by the Scripture, that such a thing can be: And the reasones refuted of all such as would call it but an imagination and Melancholicque humor.

PHILOMATHES.

Now since yee haue satisfied me nowe so fullie, concerning *Magie* or *Necromancie* I will pray you to do the like in *Sorcerie* or *Witchcraft*.

EPI. That fielde is likewise verie large: and althought in the mouthes and pennes of manie, yet fewe knowes the trueth thereof, so wel as they beleeue themselues, as I shall so shortely as I can, make you (God willing) as easelie to perceiue.

PHI. But I pray you before ye goe further, let mee interrupt you here with a shorte digression: which is, that manie can scarcely beleeue that there is such a thing as Witch-craft. Whose reasons I wil shortely alleage vnto you, that ye may satisfie me as well in that, as ye haue done in the rest. For first, whereas the Scripture seemes to prooue Witchcraft to be, by diuerse examples, and speciallie by sundrie of the same, which ye haue alleaged, it is thought by some, that these places speakes of *Magicians* and *Necromancers* onlie, & not of Witches. As in special, these wise men of *Pharaohs*, that counterfeited *Moyses* miracles, were *Magicians* say they, & not Witches: As likewise that *Pythonisse* that *Saul* consulted with: And so was *Simon Magus* in the new Testament, as that very stile importes. Secondlie, where ye would oppone the dailie practicque, & confession of so manie, that is thought likewise to be but verie melancholicque imaginations of simple rauing creatures. Thirdly, if Witches had such power of Witching of folkes to death, (as they say they haue) there had bene none left aliue long sence in the world, but they: at the least, no good or godlie person of whatsoeuer estate, coulde haue escaped their deuilrie.

EPI. Your three reasons as I take, ar grounded the first of them *negativè* vpon the Scripture: The second *affirmativè* vpon Physicke: And the thirde vpon the certaine proofe of experience. As to your first, it is most true indeede, that all these wise men of *Pharaoh* were *Magicians* of art: As likewise it appeares wel that the *Pythonisse*, with whom *Saul* consulted, was of that same profession: & so was *Simon Magus*. But yee omitted to speake of the Lawe of God, wherein are all *Magicians*, Diuines, Enchanters, Sorcerers, Witches, & whatsouer of that kinde that consultes with the Deuill, plainelie prohibited, and alike threatned against. And besides that, she who had the Spirite of *Python*, in the Actes,

Act. 16.

whose Spirite was put to silence by the Apostle, coulde be no other thing but a verie Sorcerer or Witch, if ye admit the vulgare distinction, to be in a maner true, whereof I spake in the beginning of our conference. For that spirit whereby she conquested such gaine to her Master, was not at her raising or commanding, as she pleased to appoynt, but spake by her toung, aswel publicklie, as priuatelie: Whereby she seemed to draw nearer to the sort of *Demoniakes* or possessed, if that conjunction betwixt them, had not bene of her owne consent: as it appeared by her, not being tormented therewith: And by her conquesting of such gaine to her masters (as I haue alreadie said.) As to your second reason grounded vpon Physick, in attributing their confessiones or apprehensiones, to a naturall melancholicque humour: Anie that pleases Physicallie to consider vpon the naturall humour of melancholie, according to all the Physicians, that euer writ thereupon, they shall finde that that will be ouer short a cloak to couer their knauery with: For as the humor of Melancholie in the selfe is blacke, heauie and terrene, so are the symptomes thereof, in any persones that are subject thereunto, leannes, palenes, desire of solitude: and if they come to the highest degree therof, mere folie and *Manie*: where as by the contrarie, a great number of them that euer haue bene convict or confessors of Witchcraft, as may be presently seene by manie that haue at this time confessed: they are by the contrarie, I say, some of them rich and worldly-wise, some of them fatte or corpulent in their bodies, and most part of them altogether giuen ouer to the pleasures of the flesh, continual haunting of companie, and all kind of merrines, both lawfull and vnlawfull, which are thinges directly contrary to the symptomes of Melancholie, whereof I spake, and further experience daylie proues how loath they are to confesse without torture, which witnesseth their guiltines, where by the contrary, the Melancholicques

DAEMONOLOGIE

neuer spares to bewray themselues, by their continuall discourses, feeding therby their humor in that which they thinke no crime. As to your third reason, it scarselie merites an answere. For if the deuill their master were not bridled, as the scriptures teacheth vs, suppose there were no men nor women to be his instrumentes, he could finde waies inough without anie helpe of others to wrack al mankinde: wherevnto he employes his whole study, and *goeth about like a roaring Lyon* (as PETER saith)

1. *Pet.* 5.

to that effect, but the limites of his power were set down before the foundations of the world were laid, which he hath not power in the least jote to transgresse. But beside all this, there is ouer greate a certainty to proue that they are, by the daily experience of the harmes that they do, both to men, and whatsoeuer thing men possesses, whome God will permit them to be the instrumentes, so to trouble or visite, as in my discourse of that arte, yee shall heare clearelie proued.

Chapter 2

ARGVMENT.

The Etymologie and signification of that word of Sorcerie. *The first entresse and prentishippe of them that giues themselues to that craft.*

PHILOMATHES.

Come on then I pray you, and returne where ye left.

EPI. This word of *Sorcerie* is a *Latine* worde, which is taken from casting of the lot, & therefore he that vseth it, is called *Sortiarius à sorte.* As to the word of *Witchcraft*, it is nothing but a proper name giuen in our language. The cause wherefore they were called *sortiarij*, proceeded of their practicques seeming to come of lot or chance: Such as the turning of the riddle: the knowing of the forme of prayers, or such like tokens: If a person diseased woulde liue or dye. And in generall, that name was giuen them for vsing of such charmes, and freites, as that Crafte teacheth them. Manie poynts of their craft and practicques are common betuixt the *Magicians* and them: for they serue both one Master, althought in diuerse fashions. And as I deuided the *Necromancers*, into two sorts, learned and vnlearned; so must I denie them in other two, riche and of better accompt, poore and of basser degree. These two degrees now of persones, that practises this craft, answers to the passions in them, which (I told you before) the Deuil vsed as meanes to intyse them to his seruice, for such of them as are in great miserie and pouertie, he allures to follow him, by promising vnto them greate riches, and worldlie commoditie. Such as though riche, yet burnes in a desperat desire of reuenge, hee allures them by promises, to get their turne satisfied to their hartes contentment. It is to be noted nowe, that that olde and craftie enemie of ours, assailes none, though touched with any of these two extremities, except he first finde an entresse reddy for him, either by the great ignorance of the person he deales with, ioyned with an euill

DAEMONOLOGIE

life, or else by their carelesnes and contempt of God: And finding them in an vtter despair, for one of these two former causes that I haue spoken of; he prepares the way by feeding them craftely in their humour, and filling them further and further with despaire, while he finde the time proper to discouer himself vnto them. At which time, either vpon their walking solitarie in the fieldes, or else lying pansing in their bed; but alwaies without the company of any other, he either by a voyce, or in likenesse of a man inquires of them, what troubles them: and promiseth them, a suddaine and certaine waie of remedie, vpon condition on the other parte, that they follow his advise; and do such thinges as he wil require of them: Their mindes being prepared before hand, as I haue alreadie spoken, they easelie agreed vnto that demande of his: And syne settes an other tryist, where they may meete againe. At which time, before he proceede any further with them, he first perswades them to addict themselues to his seruice: which being easely obteined, he then discouers what he is vnto them: makes them to renunce their God and *Baptisme* directlie, and giues them his marke vpon some secreit place of their bodie, which remaines soare vnhealed, while his next meeting with them, and thereafter euer insensible, how soeuer it be nipped or pricked by any, as is dailie proued, to giue them a proofe thereby, that as in that doing, hee could hurte and heale them; so all their ill and well doing thereafter, must depende vpon him. And besides that, the intollerable dolour that they feele in that place, where he hath marked them, serues to waken them, and not to let them rest, while their next meeting againe: fearing least otherwaies they might either forget him, being as new Prentises, and not well inough founded yet, in that fiendlie follie: or els remembring of that horrible promise they made him, at their last meeting, they might skunner at the same, and preasse to call it back. At their thirde meeting, he makes a shew to be carefull to performe his promises, either by teaching them waies howe to get themselues reuenged, if they be of that sort: Or els by teaching them lessons, how by moste vilde and vnlawfull meanes, they may obtaine gaine, and worldlie commoditie, if they be of the other sorte.

Chapter 3

ARGVMENT.

The Witches *actiones diuided in two partes. The actiones proper to their owne persones. Their actiones toward others. The forme of their conuentiones, and adoring of their Master.*

PHILOMATHES.

Ye haue said now inough of their initiating in that ordour. It restes then that ye discourse vpon their practises, fra they be passed Prentises: for I would faine heare what is possible to them to performe in verie deede. Although they serue a common Master with the *Necromancers*, (as I haue before saide) yet serue they him in an other forme. For as the meanes are diuerse, which allures them to these vnlawfull artes of seruing of the Deuill; so by diuerse waies vse they their practises, answering to these meanes, which first the Deuill, vsed as instrumentes in them; though al tending to one end: To wit, the enlargeing of Sathans tyrannie, and crossing of the propagation of the Kingdome of CHRIST, so farre as lyeth in the possibilitie, either of the one or other sorte, or of the Deuill their Master. For where the *Magicians*, as allured by curiositie, in the most parte of their practises, seekes principallie the satisfying of the same, and to winne to themselues a popular honoure and estimation: These Witches on the other parte, being intised ether for the desire of reuenge, or of worldly riches, their whole practises are either to hurte men and their gudes, or what they possesse, for satisfying of their cruell mindes in the former, or else by the wracke in whatsoeuer sorte, of anie whome God will permitte them to haue power off, to satisfie their greedie desire in the last poynt.

EPI. In two partes their actiones may be diuided; the actiones of their owne persones, and the actiones proceeding from them towardes anie other. And this diuision being wel vnderstood, will easilie resolue you, what is possible to them to doe. For although all

DAEMONOLOGIE

that they confesse is no lie vpon their parte, yet doubtlesly in my opinion, a part of it is not indeede, according as they take it to be: And in this I meane by the actiones of their owne persones. For as I said before, speaking of *Magie* that the Deuill illudes the senses of these schollers of his, in manie thinges, so saye I the like of these Witches.

PHI. Then I pray you, first to speake of that part of their owne persons, and syne ye may come next to their actiones towardes others.

EPI. To the effect that they may performe such seruices of their false Master, as he employes them in, the deuill as Gods Ape, counterfeites in his seruantes this seruice & forme of adoration, that God prescribed and made his seruantes to practise. For as the seruants of GOD, publicklie vses to conveene for seruing of him, so makes he them in great numbers to conveene (though publickly they dare not) for his seruice. As none conueenes to the adoration and worshipping of God, except they be marked with his scale, the Sacrament of *Baptisme*: So none serues Sathan, and conueenes to the adoring of him, that are not marked with that marke, wherof I alredy spake. As the Minister sent by God, teacheth plainely at the time of their publick conuentions, how to serue him in spirit & truth: so that vncleane spirite, in his owne person teacheth his Disciples, at the time of their conueening, how to worke all kinde of mischiefe: And craues compt of all their horrible and detestable proceedinges passed, for aduancement of his seruice. Yea, that he may the more viuelie counterfeit and scorne God, he oft times makes his slaues to conueene in these verrie places, which are destinat and ordeined for the conveening of the servantes of God (I meane by Churches). But this farre, which I haue yet said, I not onelie take it to be true in their opiniones, but euen so to be indeede. For the forme that he vsed in counterfeiting God amongst the *Gentiles*, makes me so to thinke: As God spake by his Oracles, spake he not so by his? As GOD had aswell bloudie Sacrifices, as others without bloud, had not he the like? As God had Churches sanctified to his seruice, with Altars, Priests, Sacrifices, Ceremonies and Prayers; had he not the like polluted to his seruice? As God gaue responses by *Vrim* and *Thummim*, gaue he not his responses by the intralls of beastes, by the singing of Fowles, and by their actiones in the aire? As God by visiones, dreames, and extases reueiled what was to come, and what was his will vnto his seruantes; vsed he not the like meanes to forwarne his slaues of things to come? Yea, euen as God loued cleannes, hated vice, and impuritie, & appoynted punishmentes therefore: vsed he not the like (though falselie I grant, and but in eschewing the lesse inconuenient, to draw them upon a greater) yet

dissimuled he not I say, so farre as to appoynt his Priestes to keepe their bodies cleane and vndefiled, before their asking responses of him? And feyned he not God to be a protectour of euerie vertue, and a iust reuenger of the contrarie? This reason then moues me, that as he is that same Deuill; and as craftie nowe as he was then; so wil hee not spare a pertelie in these actiones that I haue spoken of, concerning the witches persones: But further, Witches oft times confesses not only his conueening in the Church with them, but his occupying of the Pulpit: Yea, their forme of adoration, to be the kissing of his hinder partes. Which though it seeme ridiculous, yet may it likewise be true, seeing we reade that in *Calicute*, he appearing in forme of a *Goate*-bucke, hath publicklie that vn-honest homage done vnto him, by euerie one of the people: So ambitious is he, and greedie of honour (which procured his fall) that he will euen imitate God in that parte,
Exo. 33.
where it is said, that *Moyses* could see but the *hinder partes of God, for the brightnesse of his glorie*: And yet that speache is spoken but ανθρωπωπαθειαν.

Chapter 4

ARGVMENT.
What are the waies possible, wherby the witches may transport themselues to places far distant, And what ar impossible & mere illusiones of Sathan. And the reasons therof.

PHILOMATHES.
Bvt by what way say they or think ye it possible that they can com to these vnlawful cõuentiõs?

EPI. There is the thing which I esteeme their senses to be deluded in, and though they lye not in confessing of it, because they thinke it to be true, yet not to be so in substance or effect: for they saie, that by diuerse meanes they may conueene, either to the adoring of their Master, or to the putting in practise any seruice of his, committed vnto their charge: one way is natural, which is natural riding, going or sayling, at what houre their Master comes and aduertises them. And this way may be easelie beleued: an other way is some-what more strange: and yet is it possible to be true: which is by being carryed by the force of the Spirite which is their conducter, either aboue the earth or aboue the Sea swiftlie, to the place where they are to meet: which I am perswaded to be likewaies possible, in respect that as *Habakkuk* was carryed by the Angell in that forme, to the denne where *Daniell* laie;

Apocrypha of Bell and the Dragon.

so thinke I, the Deuill will be reddie to imitate God, as well in that as in other thinges: which is much more possible to him to doe, being a Spirite, then to a mighty winde, being but a naturall meteore, to transporte from one place to an other a solide bodie, as is commonlie and dailie seene in practise: But in this violent forme they cannot be carryed, but a shorte boundes, agreeing with the space that they may reteine their breath: for if it were longer, their breath could not remaine vnextinguished, their bodie being carryed in such a violent

& forceable maner, as be example: If one fall off an small height, his life is but in perrell, according to the harde or soft lighting: But if one fall from an high and stay rocke, his breath wilbe forceablie banished from the bodie, before he can win to the earth, as is oft seen by experience. And in this transporting they say themselues, that they are inuisible to anie other, except amongst themselues; which may also be possible in my opinion. For if the deuil may forme what kinde of impressiones he pleases in the aire, as I haue said before, speaking of *Magie*, why may he not far easilier thicken & obscure so the air, that is next about them by contracting it strait together, that the beames of any other mans eyes, cannot pearce thorow the same, to see them? But the third way of their comming to their conuentions, is, that where in I think them deluded: for some of them sayeth, that being transformed in the likenesse of a little beast or foule, they will come and pearce through whatsoeuer house or Church, though all ordinarie passages be closed, by whatsoeuer open, the aire may enter in at. And some sayeth, that their bodies lying stil as in an extasy, their spirits wil be rauished out of their bodies, & caried to such places. And for verefying therof, wil giue euident tokens, aswel by witnesses that haue seene their body lying senseles in the meane time, as by naming persones, whom-with they mette, and giuing tokens what purpose was amongst them, whome otherwaies they could not haue knowen: for this forme of journeing, they affirme to vse most, when they are transported from one Countrie to another.

PHI. Surelie I long to heare your owne opinion of this: For they are like old wiues trattles about the fire. The reasons that moues me to thinke that these are meere illusiones, ar these. First for them that are transformed in likenes of beastes or foules, can enter through so narrow passages, although I may easelie beleeue that the Deuill coulde by his woorkemanshippe vpon the aire, make them appeare to be in such formes, either to themselues or to others: Yet how he can contract a solide bodie within so little roome, I thinke it is directlie contrarie to it selfe, for to be made so little, and yet not diminished: To be so straitlie drawen together, and yet feele no paine; I thinke it is so contrarie to the qualitie of a naturall bodie, and so like to the little transubstantiat god in the *Papistes Masse*, that I can neuer beleeue it. So to haue a quantitie, is so proper to a solide bodie, that as all Philosophers conclude, it cannot be any more without one, then a spirite can haue one. For when PETER *came out of the prison,*
    Act. 12.

DAEMONOLOGIE

and the doores all locked: It was not by any contracting of his bodie in so little roome: but by the giuing place of the dore, though vn-espyed by the Gaylors. And yet is there no comparison, when this is done, betuixt the power of God, and of the Deuill. As to their forme of extasie and spirituall transporting, it is certaine the soules going out of the bodie, is the onely difinition of naturall death: and who are once dead, God forbid wee should thinke that it should lie in the power of all the Deuils in Hell, to restore them to their life againe: Although he can put his owne spirite in a dead bodie, which the *Necromancers* commonlie practise, as yee haue harde. For that is the office properly belonging to God; and besides that, the soule once parting from the bodie, cannot wander anie longer in the worlde, but to the owne resting place must it goe immediatlie, abiding the conjunction of the bodie againe, at the latter daie. And what CHRIST or the Prophets did miraculouslie in this case, it cannot in no Christian mans opinion be maid common with the Deuill. As for anie tokens that they giue for proouing of this, it is verie possible to the Deuils craft, to perswade them to these meanes. For he being a spirite, may hee not so rauishe their thoughtes, and dull their sences, that their bodie lying as dead, hee may object to their spirites as it were in a dreame, & (as the Poets write of *Morpheus*) represente such formes of persones, of places, and other circumstances, as he pleases to illude them with? Yea, that he maie deceiue them with the greater efficacie, may hee not at that same instant, by fellow angelles of his, illude such other persones so in that same fashion, whome with he makes them to beleeue that they mette; that all their reportes and tokens, though seuerallie examined, may euerie one agree with an other. And that whatsoeuer actiones, either in hurting men or beasts: or whatsoeuer other thing that they falselie imagine, at that time to haue done, may by himselfe or his marrowes, at that same time be done indeede; so as if they would giue for a token of their being rauished at the death of such a person within so shorte space thereafter, whom they beleeue to haue poysoned, or witched at that instante, might hee not at that same houre, haue smitten that same person by the permission of GOD, to the farther deceiuing of them, and to mooue others to beleeue them? And this is surelie the likeliest way, and most according to reason, which my judgement can finde out in this, and whatsoeuer vther vnnaturall poyntes of their confession. And by these meanes shall we saill surelie, betuixt *Charybdis* and *Scylla*, in eschewing the not beleeuing of them altogether on the one part, least that drawe vs to the errour that there is no Witches: and on the other parte in beleeuing of it, make vs to eschew the falling into innumerable absurdities, both monstruouslie against all Theologie diuine, and Philosophie humaine.

Chapter 5

ARGVMENT.

Witches actiones towardes others. Why there are more women of that craft nor men? What thinges are possible to them to effectuate by the power of their master. The reasons thereof. What is the surest remedie of the harmes done by them.

PHILOMATHES.

Forsooth your opinion in this, seemes to carrie most reason with it, and sence yee haue ended, then the actions belonging properly to their owne persones: say forwarde now to their actiones vsed towardes others.

EPI. In their actiones vsed towardes others, three thinges ought to be considered: First the maner of their consulting thereupon: Next their part as instrumentes: And last their masters parte, who puts the same in execution. As to their consultationes thereupon, they vse them oftest in the Churches, where they conveene for adoring: at what time their master enquiring at them what they would be at: euerie one of them propones vnto him, what wicked turne they would haue done, either for obteining of riches, or for reuenging them vpon anie whome they haue malice at: who granting their demande, as no doubt willinglie he wil, since it is to doe euill, he teacheth them the means, wherby they may do the same. As for little trifling turnes that women haue ado with, he causeth them to ioynt dead corpses, & to make powders thereof, mixing such other thinges there amongst, as he giues vnto them.

PHI. But before yee goe further, permit mee I pray you to interrupt you one worde, which yee haue put mee in memorie of, by speaking of Women. What can be the cause that there are twentie women giuen to that craft, where ther is one man?

EPI. The reason is easie, for as that sexe is frailer then man is, so is it easier to be intrapped in these grosse snares of the Deuill, as was ouer well proued to be true, by the Serpents deceiuing of *Eua* at the beginning, which makes him the homelier with that sexe sensine.

PHI. Returne now where ye left.

EPI. To some others at these times hee teacheth how to make Pictures of waxe or clay: That by the rosting thereof, the persones that they beare the name of, may be continuallie melted or dryed awaie by continuall sicknesse. To some hee giues such stones or poulders, as will helpe to cure or cast on diseases: And to some he teacheth kindes of vncouthe poysons, which Mediciners vnderstandes not (for he is farre cunningner then man in the knowledge of all the occult proprieties of nature) not that anie of these meanes which hee teacheth them (except the poysons which are composed of thinges naturall) can of them selues helpe any thing to these turnes, that they are employed in, but onelie being Gods Ape, as well in that, as in all other thinges. Even as God by his Sacramentes which are earthlie of themselues workes a heauenlie effect, though no waies by any cooperation in them: And

Iohn. 9.

as CHRIST by clay & spettle wrought together, *opened the eies of the blynd man,* suppose there was no vertue in that which he outwardlie applyed, so the Deuill will haue his out-warde meanes to be shewes as it were of his doing, which hath no part of cooperation in his turnes with him, how farre that euer the ignorantes be abused in the contrarie. And as to the effectes of these two former partes, to wit, the consultationes and the outward meanes, they are so wounderfull as I dare not allege anie of them, without ioyning a sufficient reason of the possibilitie thereof. For leauing all the small trifles among wiues, and to speake of the principall poyntes of their craft. For the common trifles thereof, they can do without conuerting well inough by themselues: These principall poyntes I say are these: They can make men or women to loue or hate other, which may be verie possible to the Deuil to effectual, seing he being a subtile spirite, knowes well inough how to perswade the corrupted affection of them whom God will permit him so to deale with: They can lay the siknesse of one vpon an other, which likewise is verie possible vnto him: For since by Gods permission, he layed siknesse vpon IOB, why may he not farre easilier lay it vpon any other: For as an old practisian, he knowes well inough what humor domines most in anie of vs, and as a spirite hee can subtillie walken vp the same, making it peccant, or to abounde, as he thinkes meete for troubling of vs, when God will so permit him. And for the taking off of it, no doubt he will be glad to reliue such of present paine, as he may thinke by these meanes to perswade to bee catched in his euerlasting snares and fetters. They can be-witch and take the life of men or women, by rosting of the Pictures, as I spake of before, which likewise is verie possible to their Master to performe, for although, (as I saide before) that instrumente of waxe haue no vertue in that turne doing, yet may hee not verie well euen by that same measure that his conjured slaues meltes that waxe at the fire, may he not I say at these same times, subtilie as a spirite so weaken and scatter the spirites of life of the patient, as may make him on th'one part, for faintnesse to sweate out the humour of his bodie: And on the other parte, for the not concurrence of these spirites, which causes his digestion, so debilitat his stomak, that his humour radicall continually, sweating out on the one parte, and no new good suck being put in the place thereof, for lack of digestion on the other, hee at last shall vanish awaie, euen as his picture will doe at the fire. And that knauish and cunning woorkeman, by troubling him onely at some times, makes a proportion so neare betuixt the woorking of the one and the other, that both shall ende as it were at one time. They can rayse stormes and tempestes in the aire, either vpon Sea or land, though not vniuersally, but in such a particular place and prescribed boundes, as God will permitte them so to trouble: Which likewise is verie easie to be discerned from anie other naturall tempestes that are meteores, in respect of the suddaine and violent raising thereof, together with the short induring of the same. And this is likewise verie possible to their master to do, he hauing such affinitie with the aire as being a spirite, and hauing such power of the forming and moouing thereof, as ye haue heard me alreadie declare: For in the Scripture, that stile of *the Prince of the aire*

Ephes. 2.

is giuen vnto him. They can make folkes to becom phrenticque or Maniacque, which likewise is very possible to their master to do, sence they are but naturall sicknesses: and so he may lay on these kindes, aswell as anie others. They can make spirites either to follow and trouble persones, or haunt certaine houses, and affraie oftentimes the inhabitantes: as hath bene knowen to be done by our Witches at this time. And likewise they can make some to be possessed with spirites, & so to becom verie Dæmoniacques: and this last sorte is verie possible likewise to the Deuill their Master to do, since he may easilie send his owne angells to trouble in what forme he pleases, any whom God wil permit him so to vse.

PHI. But will God permit these wicked instrumentes by the power of the Deuill their master, to trouble by anie of these meanes, anie that beleeues in him?

EPI. No doubt, for there are three kinde of folkes whom God will permit so to be tempted or troubled; the wicked for their horrible sinnes, to punish them in the like measure; The godlie that are sleeping in anie great sinnes or infirmities and weakenesse in faith, to waken them vp the faster by such an vncouth forme: and euen some of the best, that their patience may bee tryed before the world, as IOBS was. For why may not God vse anie kinde of extraordinarie punishment, when it pleases him; as well as the ordinarie roddes of sicknesse or other aduersities. PHI. Who then may be free from these Deuilish practises?

EPI. No man ought to presume so far as to promise anie impunitie to himselfe: for God hath before all beginninges preordinated aswell the particular sortes of Plagues as of benefites for euerie man, which in the owne time he ordaines them to be visited with, & yet ought we not to be the more affrayde for that, of any thing that the Deuill and his wicked instrumentes can do against vs: For we dailie fight against the Deuill in a hundreth other waies: And therefore as a valiant Captaine, affraies no more being at the combat, nor stayes from his purpose for the rummishing shot of a Cannon, nor the small clack of a Pistolet: suppose he be not certaine what may light vpon him; Euen so ought we boldlie to goe forwarde in fighting against the Dcuill without anie greater terrour, for these his rarest weapons, nor for the ordinarie whereof wee haue daily the proofe.

PHI. Is it not lawfull then by the helpe of some other Witche to cure the disease that is casten on by that craft?

EPI. No waies lawfull: For I gaue you the reason thereof in that axiome of Theologie, which was the last wordes I spake of *Magie*.

PHI. How then may these diseases be lawfullie cured?

EPI. Onelie by earnest prayer to GOD, by amendement of their liues, and by sharp persewing euerie one, according to his calling of these instrumentes of Sathan, whose punishment to the death will be a salutarie sacrifice for the patient. And this is not onely the lawfull way, but likewise the most sure: For by the Deuils meanes, *can neuer the Deuill be casten out,*

Mark. 3.

as Christ sayeth. And when such a cure is vsed, it may wel serue for a shorte time, but at the last, it will doubtleslie tend to the vtter perdition of the patient, both in bodie and soule.

Chapter 6

ARGVMENT.

What sorte of folkes are least or most subiect to receiue harme by Witchcraft. What power they haue to harme the Magistrate, and vpon what respectes they haue any power in prison: And to what end may or will the Deuill appeare to them therein. Vpon what respectes the Deuill appeires in sundry shapes to sundry of them at any time.

PHILOMATHES.

Bvt who dare take vpon him to punish them, if no man can be sure to be free from their vnnaturall inuasiones?

EPI. We ought not the more of that restraine from vertue, that the way wherby we climbe thereunto be straight and perilous. But besides that, as there is no kinde of persones so subject to receiue harme of them, as these that are of infirme and weake faith (which is the best buckler against such inuasiones:) so haue they so smal power ouer none, as ouer such as zealouslie and earnestlie persewes them, without sparing for anie worldlie respect.

PHI. Then they are like the Pest, which smites these sickarest, that flies it farthest, and apprehends deepliest the perrell thereof.

EPI. It is euen so with them: For neither is it able to them to vse anie false cure vpon a patient, except the patient first beleeue in their power, and so hazard the tinsell of his owne soule, nor yet can they haue lesse power to hurte anie, nor such as contemnes most their doinges, so being it comes of faith, and not of anie vaine arrogancie in themselues.

PHI. But what is their power against the Magistrate?

EPI. Lesse or greater, according as he deales with them. For if he be slouthfull towardes them, God is verie able to make them instrumentes to waken & punish his slouth. But if he be the contrarie, he according to the iust law of God, and allowable law of all Nationes, will be diligent in examining and punishing of them: GOD will not permit their master to trouble or hinder so good a woorke.

PHI. But fra they be once in handes and firmance, haue they anie further power in their craft?

EPI. That is according to the forme of their detention. If they be but apprehended and deteined by anie priuate person, vpon other priuate respectes, their power no doubt either in escaping, or in doing hurte, is no lesse nor euer it was before. But if on the other parte, their apprehending and detention be by the lawfull Magistrate, vpon the iust respectes of their guiltinesse in that craft, their power is then no greater then before that euer they medled with their master. For where God beginnes iustlie to strike by his lawfull Lieutennentes, it is not in the Deuilles power to defraude or bereaue him of the office, or effect of his powerfull and reuenging Scepter.

PHI. But will neuer their master come to visite them, fra they be once apprehended and put in firmance?

EPI. That is according to the estaite that these miserable wretches are in: For if they be obstinate in still denying, he will not spare, when he findes time to speake with them, either if he finde them in anie comfort, to fill them more and more with the vaine hope of some maner of reliefe: or else if hee finde them in a deepe dispaire, by all meanes to augment the same, and to perswade them by some extraordinarie meanes to put themselues downe, which verie commonlie they doe. But if they be penitent and confesse, God will not permit him to trouble them anie more with his presence and allurementes.

PHI. It is not good vsing his counsell I see then. But I woulde earnestlie know when he appeares to them in Prison, what formes vses he then to take?

EPI. Diuers formes, euen as he vses to do at other times vnto them. For as I told you, speking of *Magie*, he appeares to that kinde of craftes-men ordinarily in an forme, according as they agree vpon it amongst themselues: Or if they be but prentises, according to the qualitie of their circles or conjurationes: Yet to these capped creatures, he appeares as he pleases, and as he findes meetest for their humors. For euen at their publick conuentiones, he appeares to diuers of them in diuers formes, as we haue found by the difference of their confessiones in that point: For he deluding them with vaine impressiones in the aire, makes himselfe to seeme more terrible to the grosser sorte, that they maie thereby be moued to feare and reuerence him the more: And les monstrous and vncouthlike againe to the

DAEMONOLOGIE

craftier sorte, least otherwaies they might sturre and skunner at his vglinesse.

PHI. How can he then be felt, as they confesse they haue done him, if his bodie be but of aire?

EPI. I heare little of that amongst their confessiones, yet may he make himselfe palpable, either by assuming any dead bodie, and vsing the ministrie thereof, or else by deluding as wel their sence of feeling as seeing; which is not impossible to him to doe, since all our senses, as we are so weake, and euen by ordinarie sicknesses will be often times deluded.

PHI. But I would speere one worde further yet, concerning his appearing to them in prison, which is this. May any other that chances to be present at that time in the prison, see him as well as they. EPI. Some-times they will, and some-times not, as it pleases God.

Chapter 7

ARGVMENT.

Two formes of the deuils visible conuersing in the earth, with the reasones wherefore the one of them was communest in the time of Papistrie: And the other sensine. Those that denies the power of the Deuill, denies the power of God, and are guiltie of the errour of the Sadduces.

PHILOMATHES.

Hath the Deuill then power to appeare to any other, except to such as are his sworne disciples: especially since al Oracles, & such like kinds of illusiones were taken awaie and abolished by the cumming of CHRIST?

EPI. Although it be true indeede, that the brightnesse of the Gospell at his cumming, scaled the cloudes of all these grosse errors in the Gentilisme: yet that these abusing spirites, ceases not sensine at sometimes to appeare, dailie experience teaches vs. Indeede this difference is to be marked betwixt the formes of Sathans conuersing visiblie in the world. For of two different formes thereof, the one of them by the spreading of the Euangell, and conquest of the white horse, in the sixt Chapter of the Reuelation, is much hindred and become rarer there through. This his appearing to any Christians, troubling of them outwardly, or possessing of them constraynedly. The other of them is become communer and more vsed sensine, I meane by their vnlawfull artes, whereupon our whole purpose hath bene. This we finde by experience in this Ile to be true. For as we know, moe Ghostes and spirites were seene, nor tongue can tell, in the time of blinde *Papistrie* in these Countries, where now by the contrarie, a man shall scarcely all his time here once of such things. And yet were these vnlawfull artes farre rarer at that time: and neuer were so much harde of, nor so rife as they are now.

PHI. What should be the cause of that?

DAEMONOLOGIE

EPI. The diuerse nature of our sinnes procures at the Iustice of God, diuerse sortes of punishments answering thereunto. And therefore as in the time of *Papistrie*, our fathers erring grosselie, & through ignorance, that mist of errours ouershaddowed the Deuill to walke the more familiarlie amongst them: And as it were by barnelie and affraying terroures, to mocke and accuse their barnelie erroures. By the contrarie, we now being sounde of Religion, and in our life rebelling to our profession, God iustlie by that sinne of rebellion, as *Samuel* calleth it, accuseth our life so wilfullie fighting against our profession.

PHI. Since yee are entred now to speake of the appearing of spirites: I would be glad to heare your opinion in that matter. For manie denies that anie such spirites can appeare in these daies as I haue said.

EPI. Doubtleslie who denyeth the power of the Deuill, woulde likewise denie the power of God, if they could for shame. For since the Deuill is the verie contrarie opposite to God, there can be no better way to know God, then by the contrarie; as by the ones power (though a creature) to admire the power of the great Creator: by the falshood of the one to considder the trueth of the other, by the injustice of the one, to considder the Iustice of the other: And by the cruelty of the one, to considder the mercifulnesse of the other: And so foorth in all the rest of the essence of God, and qualities of the Deuill. But I feare indeede, there be ouer many *Sadduces* in this worlde, that denies all kindes of spirites: For convicting of whose errour, there is cause inough if there were no more, that God should permit at sometimes spirits visiblie to kyith.

Thirde Booke

ARGVMENT.
The description of all these kindes of Spirites that troubles men or women. The conclusion of the whole Dialogue.

Chapter 1

ARGVMENT.

The diuision of spirites in foure principall kindes. The description of the first kinde of them, called Spectra & vmbræ mortuorum. What is the best way to be free of their trouble.

PHILOMATHES.

I pray you now then go forward in telling what ye thinke fabulous, or may be trowed in that case.

EPI. That kinde of the Deuils conuersing in the earth, may be diuided in foure different kindes, whereby he affrayeth and troubleth the bodies of men: For of the abusing of the soule, I haue spoken alreadie. The first is, where spirites troubles some houses or solitarie places: The second, where spirites followes vpon certaine persones, and at diuers houres troubles them: The thirde, when they enter within them and possesse them: The fourth is these kinde of spirites that are called vulgarlie the Fayrie. Of the three former kindes, ye harde alreadie, how they may artificiallie be made by Witch-craft to trouble folke: Now it restes to speake of their naturall comming as it were, and not raysed by Witch-craft. But generally I must for-warne you of one thing before I enter in this purpose: that is, that although in my discourseing of them, I deuyde them in diuers kindes, yee must notwithstanding there of note my Phrase of speaking in that: For doubtleslie they are in effect, but all one kinde of spirites, who for abusing the more of mankinde, takes on these sundrie shapes, and vses diuerse formes of out-ward actiones, as if some were of nature better then other. Nowe I returne to my purpose: As to the first kinde of these spirites, that were called by the auncients by diuers names, according as their actions were. For if they were spirites that haunted some houses, by appearing in diuers and horrible formes, and making greate dinne: they were called *Lemures* or *Spectra*. If they appeared in likenesse of anie defunct to some friends of his, they wer called *vmbræ mortuorum*: And so innumerable stiles they got, according to their actiones, as I haue said alreadie. As we see by experience, how manie stiles they haue given them in our language in the like maner: Of the appearing of these spirites, wee are certified by the Scriptures, where the Prophet ESAY 13.

Esay. 13. *Iere*. 50.

and 34. cap. threatning the destruction of *Babell* and *Edom*: declares, that it shal not onlie be wracked, but shall become so greate a solitude, as it shall be the habitackle of Howlettes, and of ZIIM and IIM, which are the proper Hebrewe names for these Spirites. The cause whie they haunte solitarie places, it is by reason, that they may affraie and brangle the more the faith of suche as them alone hauntes such places. For our nature is such, as in companies wee are not so soone mooued to anie such kinde of feare, as being solitare, which the Deuill knowing well inough, hee will not therefore assaile vs but when we are weake: And besides that, GOD will not permit him so to dishonour the societies and companies of Christians, as in publicke times and places to walke visiblie amongst them. On the other parte, when he troubles certaine houses that are dwelt in, it is a sure token either of grosse ignorance, or of some grosse and slanderous sinnes amongst the inhabitantes thereof: which God by that extraordinarie rod punishes.

PHI. But by what way or passage can these Spirites enter in these houses, seeing they alledge that they will enter, Doore and Window being steiked?

EPI. They will choose the passage for their entresse, according to the forme that they are in at that time. For if they haue assumed a deade bodie, whereinto they lodge themselues, they can easely inough open without dinne anie Doore or Window, and enter in thereat. And if they enter as a spirite onelie, anie place where the aire may come in at, is large inough an entrie for them: For as I said before, a spirite can occupie no quantitie.

PHI. And will God then permit these wicked spirites to trouble the reste of a dead bodie, before the resurrection thereof? Or if he will so, I thinke it should be of the reprobate onely.

EPI. What more is the reste troubled of a dead bodie, when the Deuill carryes it out of the Graue to serue his turne for a space, nor when the Witches takes it vp and joyntes it, or when as Swine wortes vppe the graues? The rest of them that the Scripture speakes of, is not meaned by a locall remaining continuallie in one place, but by their resting from their trauelles and miseries of this worlde, while their latter conjunction againe with the soule at that time to receaue full

DAEMONOLOGIE

glorie in both. And that the Deuill may vse aswell the ministrie of the bodies of the faithfull in these cases, as of the vn-faithfull, there is no inconvenient; for his haunting with their bodies after they are deade, can no-waies defyle them: In respect of the soules absence. And for anie dishonour it can be vnto them, by what reason can it be greater, then the hanging, heading, or many such shameful deaths, that good men will suffer? for there is nothing in the bodies of the faithfull, more worthie of honour, or freer from corruption by nature, nor in these of the vnfaithful, while time they be purged and glorified in the latter daie, as is dailie seene by the vilde diseases and corruptions, that the bodies of the faythfull are subject vnto, as yee will see clearelie proued, when I speake of the possessed and Dæmoniacques.

PHI. Yet there are sundrie that affirmes to haue haunted such places, where these spirites are alleaged to be: And coulde neuer heare nor see anie thing.

EPI. I thinke well: For that is onelie reserued to the secreete knowledge of God, whom he wil permit to see such thinges, and whome not.

PHI. But where these spirites hauntes and troubles anie houses, what is the best waie to banishe them?

EPI. By two meanes may onelie the remeid of such things be procured: The one is ardent prayer to God, both of these persones that are troubled with them, and of that Church whereof they are. The other is the purging of themselues by amendement of life from such sinnes, as haue procured that extraordinarie plague.

PHI. And what meanes then these kindes of spirites, when they appeare in the shaddow of a person newlie dead, or to die, to his friendes?

EPI. When they appeare vpon that occasion, they are called Wraithes in our language. Amongst the *Gentiles* the Deuill vsed that much, to make them beleeue that it was some good spirite that appeared to them then, ether to forewarne them of the death of their friend; or else to discouer vnto them, the will of the defunct, or what was the way of his slauchter, as is written in the booke of the histories Prodigious. And this way hee easelie deceiued the *Gentiles*, because they knew not God: And to that same effect is it, that he now appeares in that maner to some ignorant Christians. For he dare not so illude anie that knoweth that, neither can the spirite of the defunct returne to his friend, or yet an Angell vse such formes.

PHI. And are not our war-woolfes one sorte of these spirits also, that hauntes and troubles some houses or dwelling places?

EPI. There hath indeede bene an old opinion of such like thinges; For by the *Greekes* they were called λυκανθρωποι which signifieth men-woolfes. But to tell you simplie my opinion in this, if anie such thing hath bene, I take it to haue proceeded but of a naturall super-abundance of Melancholie, which as wee reade, that it hath made some thinke themselues Pitchers, and some horses, and some one kinde of beast or other: So suppose I that it hath so viciat the imagination and memorie of some, as *per lucida interualla*, it hath so highlie occupyed them, that they haue thought themselues verrie Woolfes indeede at these times: and so haue counterfeited their actiones in goeing on their handes and feete, preassing to deuoure women and barnes, fighting and snatching with all the towne dogges, and in vsing such like other bruitish actiones, and so to become beastes by a strong apprehension,

Dan. 4.

as *Nebucad-netzar* was seuen yeares: but as to their hauing and hyding of their hard & schellie sloughes, I take that to be but eiked, by vncertaine report, the author of all lyes.

Chapter 2

ARGVMENT.

The description of the next two kindes of Spirites, whereof the one followes outwardlie, the other possesses inwardlie the persones that they trouble. That since all Prophecies and visiones are nowe ceased, all spirites that appeares in these formes are euill.

PHILOMATHES.

Come forward now to the reste of these kindes of spirites.

EPI. As to the next two kindes, that is, either these that outwardlie troubles and followes some persones, or else inwardlie possesses them: I will conjoyne them in one, because aswel the causes ar alike in the persons that they are permitted to trouble: as also the waies whereby they may be remedied and cured.

PHI. What kinde of persones are they that vses to be so troubled?

EPI. Two kindes in speciall: Either such as being guiltie of greeuous offences, God punishes by that horrible kinde of scourdge, or else being persones of the beste nature peraduenture, that yee shall finde in all the Countrie about them, GOD permittes them to be troubled in that sort, for the tryall of their patience, and wakening vp of their zeale, for admonishing of the beholders, not to truste ouer much in themselues, since they are made of no better stuffe, and peraduenture blotted with no smaller sinnes (as CHRIST saide,

Luc. 13.

speaking of them vppon whome the Towre in *Siloam* fell:) And for giuing likewise to the spectators, matter to prayse GOD, that they meriting no better, are yet spared from being corrected in that fearefull forme.

PHI. These are good reasones for the parte of GOD, which apparantlie mooues him so to permit the Deuill to trouble such persones. But since the Deuil hath euer a contrarie respecte in all the actiones that GOD employes him in: which is I pray you the end and mark he shoots at in this turne?

EPI. It is to obtaine one of two thinges thereby, if he may: The one is the tinsell of their life, by inducing them to such perrilous places at such time as he either followes or possesses them, which may procure the same: And such like, so farre as GOD will permit him, by tormenting them to weaken their bodie, and caste them in incurable diseases. The other thinge that hee preases to obteine by troubling of them, is the tinsell of their soule, by intising them to mistruste and blaspheme God: Either for the intollerablenesse of their tormentes, as he assayed to haue done with IOB;

Iob. i.

or else for his promising vnto them to leaue the troubling of them, incase they would so do, as is knowen by experience at this same time by the confession of a young one that was so troubled.

PHI. Since ye haue spoken now of both these kindes of spirites comprehending them in one: I must nowe goe backe againe in speering some questions of euerie one of these kindes in speciall. And first for these that followes certaine persones, yee know that there are two sortes of them: One sorte that troubles and tormentes the persones that they haunt with: An other sort that are seruiceable vnto them in all kinde of their necessaries, and omittes neuer to forwarne them of anie suddaine perrell that they are to be in. And so in this case, I would vnderstande whither both these sortes be but wicked and damned spirites: Or if the last sorte be rather Angells, (as should appeare by their actiones) sent by God to assist such as he speciallie fauoures. For it is written in the Scriptures,

Gen. 32. 1. *Kin.* 6. *Psal.* 34.

that *God sendes Legions of Angels to guarde and watch ouer his elect.*

EPI. I know well inough where fra that errour which ye alleage hath proceeded: For it was the ignorant *Gentiles* that were the fountaine thereof. Who for that they knew not God, they forged in their owne imaginationes, euery man to be still accompanied with two spirites, whereof they called the one *genius bonus*, the other *genius malus*: the Greekes called them ευδαιμονα & κακοδαιμονα: wherof the former they saide, perswaded him to all the good he did: the other entised him to all the euill. But praised be God we that are christians, & walks not amongst the *Cymmerian* conjectures of man, knowes well inough, that it is the good spirite of God onely, who is the fountain of all goodnes, that perswads vs to the thinking or doing of any good: and

DAEMONOLOGIE

that it is our corrupted fleshe and Sathan, that intiseth vs to the contrarie. And yet the Deuill for confirming in the heades of ignoraunt Christians, that errour first mainteined among the Gentiles, he whiles among the first kinde of spirits that I speak of, appeared in time of *Papistrie* and blindnesse, and haunted diuers houses, without doing any euill, but doing as it were necessarie turnes vp and down the house: and this spirit they called *Brownie* in our language, who appeared like a rough-man: yea, some were so blinded, as to beleeue that their house was all the sonsier, as they called it, that such spirites resorted there.

PHI. But since the Deuils intention in all his actions, is euer to do euill, what euill was there in that forme of doing, since their actions outwardly were good.

EPI. Was it not euill inough to deceiue simple ignorantes, in making them to take him for an Angell of light, and so to account of Gods enemie, as of their particular friend: where by the contrarie, all we that are Christians, ought assuredly to know that since the comming of Christ in the flesh, and establishing of his Church by the Apostles, all miracles, visions, prophecies, & appearances of Angels or good spirites are ceased. Which serued onely for the first sowing of faith, & planting of the Church. Where now the Church being established, and the white Horse whereof I spake before, hauing made his conqueste, the Lawe and Prophets are thought sufficient to serue vs, or make vs inexcusable,

Luk. 16.

as Christ saith in his parable of *Lazarus* and the riche man.

Chapter 3

ARGVMENT.

The description of a particular sort of that kind of following spirites, called Incubi and Succubi: And what is the reason wherefore these kindes of spirites hauntes most the Northeme and barbarous partes of the world.

PHILOMATHES.

The next question that I would speere, is likewise concerning this first of these two kindes of spirites that ye haue conjoyned: and it is this; ye knowe how it is commonly written and reported, that amongst the rest of the sortes of spirites that followes certaine persons, there is one more monstrous nor al the rest: in respect as it is alleaged, they converse naturally with them whom they trouble and hauntes with: and therefore I would knowe in two thinges your opinion herein: First if suche a thing can be: and next if it be: whether there be a difference of sexes amongst these spirites or not.

EPI. That abhominable kinde of the Deuils abusing of men or women, was called of old, *Incubi* and *Succubi*, according to the difference of the sexes that they conuersed with. By two meanes this great kinde of abuse might possibly be performed: The one, when the Deuill onelie as a spirite, and stealing out the sperme of a dead bodie, abuses them that way, they not graithlie seeing anie shape or feeling anie thing, but that which he so conuayes in that part: As we reade of a Monasterie of Nunnes which were burnt for their being that way abused. The other meane is when he borrowes a dead bodie and so visiblie, and as it seemes vnto them naturallie as a man converses with them. But it is to be noted, that in whatsoeuer way he vseth it, that sperme seemes intollerably cold to the person abused. For if he steale out the nature of a quick person, it cannot be so quicklie carryed, but it will both tine the strength and heate by the way, which it could neuer haue had for lacke of agitation, which in the time of procreation is the procurer & wakener vp of these two natural qualities. And if he occupying the dead bodie as his lodging expell the same out thereof in the dewe time, it must likewise be colde by the participation with the qualities of the dead bodie whereout of it comes. And whereas yee inquire if these spirites be diuided in sexes or not, I thinke the rules of Philosophie may easelie resolue a man of the contrarie: For it is a sure principle of that arte, that nothing can be diuided in sexes, except such liuing bodies as must haue a naturall seede to genere by. But we know spirites hath no seede proper to themselues, nor yet can they gender one with an other.

PHI. How is it then that they say sundrie monsters haue bene gotten by that way.

EPI. These tales are nothing but *Aniles fabulæ*. For that they haue no nature of their owne, I haue shewed you alreadie. And that the cold nature of a dead bodie, can woorke nothing in generation, it is more nor plaine, as being already dead of it selfe as well as the rest of the bodie is, wanting the naturall heate, and such other naturall operation, as is necessarie for woorking that effect, and incase such a thing were possible (which were all utterly against all the rules of nature) it would breed no monster, but onely such a naturall of-spring, as would haue cummed betuixt that man or woman and that other abused person, in-case they both being aliue had had a doe with other. For the Deuilles parte therein, is but the naked carrying or expelling of that substance: And so it coulde not participate with no qualitie of the same. Indeede, it is possible to the craft of the Deuill to make a womans bellie to swel after he hath that way abused her, which he may do, either by steiring vp her own humor, or by herbes, as we see beggars daily doe. And when the time of her deliuery should come to make her thoil great doloures, like vnto that naturall course, and then subtillie to slippe in the Mid-wiues handes, stockes, stones, or some monstruous barne brought from some other place, but this is more reported and gessed at by others, nor beleeued by me.

PHI. But what is the cause that this kinde of abuse is thought to be most common in such wild partes of the worlde, as *Lap-land*, and *Fin-land*, or in our North Iles of *Orknay* and *Schet-land*.

EPI. Because where the Deuill findes greatest ignorance and barbaritie, there assayles he grosseliest, as I gaue you the reason wherefore there was moe Witches of women kinde nor men.

PHI. Can anie be so vnhappie as to giue their willing consent to the Deuilles vilde abusing them in this forme.

EPI. Yea, some of the Witches haue confessed, that he hath perswaded them to giue their willing consent thereunto, that he may thereby haue them feltred the sikarer in his snares; But as the other compelled sorte is to be pittied and prayed for, so is this most highlie to be punished and detested.

PHI. It is not the thing which we cal the *Mare*, which takes folkes sleeping in their bedds, a kinde of these spirites, whereof ye are speaking?

EPI. No, that is but a naturall sicknes, which the Mediciners hath giuen that name of *Incubus* vnto *ab incubando*, because it being a thicke fleume, falling into our breast vpon the harte, while we are sleeping, intercludes so our vitall spirites, and takes all power from vs, as maks vs think that there were some vnnaturall burden or spirite, lying vpon vs and holding vs downe.

Chapter 4

ARGVMENT.
The description of the Dæmoniackes & possessed. By what reason the Papistes may haue power to cure them.
PHILOMATHES.
Wel, I haue told you now all my doubts, and ye haue satisfied me therein, concerning the first of these two kindes of spirites that ye haue conjoyned. Now I am to inquire onely two thinges at you concerning the last kinde, I meane the Dæmoniackes. The first is, whereby shal these possessed folks be discerned fra them that ar trubled with a natural Phrensie or Manie. The next is, how can it be that they can be remedied by the Papistes Church, whome wee counting as Hereticques,
Mat. 12. *Mark.* 3.
it should appeare that one Deuil should not cast out an other, for then would *his kingdome be diuided in it selfe*, as CHRIST said.
EPI. As to your first question; there are diuers symptomes, whereby that heauie trouble may be discerned from a naturall sickenesse, and speciallie three, omitting the diuers vaine signes that the *Papistes* attributes vnto it: Such as the raging at holie water, their fleeing a back from the Croce, their not abiding the hearing of God named, and innumerable such like vaine thinges that were alike fashious and feckles to recite. But to come to these three symptomes then, whereof I spake, I account the one of them to be the incredible strength of the possessed creature, which will farre exceede the strength of six of the wightest and wodest of any other men that are not so troubled. The next is the boldning vp so far of the patients breast and bellie, with such an vnnaturall sturring and vehement agitation within them: And such an ironie hardnes of his sinnowes so stiffelie bended out, that it were not possible to prick out as it were the skinne of anie other person so far: so mightely works the Deuil in all

the members and senses of his body, he being locallie within the same, suppose of his soule and affectiones thereof, hee haue no more power then of any other mans. The last is, the speaking of sundrie languages, which the patient is knowen by them that were acquainte with him neuer to haue learned, and that with an vncouth and hollowe voice, and al the time of his speaking, a greater motion being in his breast then in his mouth. But fra this last symptome is excepted such, as are altogether in the time of their possessing bereft of al their senses being possessed with a dumme and blynde spirite, whereof Christ releiued one, in the 12. of *Mathew*. And as to your next demande, it is first to be doubted if the *Papistes* or anie not professing the the onelie true Religion, can relieue anie of that trouble. And next, in-case they can, vpon what respectes it is possible vnto them. As to the former vpon two reasons, it is grounded: first that it is knowen so manie of them to bee counterfite, which wyle the Clergie inuentes for confirming of their rotten Religion. The next is, that by experience we finde that few, who are possessed indeede, are fullie cured by them: but rather the Deuill is content to release the bodelie hurting of them, for a shorte space, thereby to obteine the perpetual hurt of the soules of so many that by these false miracles may be induced or confirmed in the profession of that erroneous Religion: euen as I told you before that he doth in the false cures, or casting off of diseases by Witches. As to the other part of the argument in-case they can, which rather (with reuerence of the learned thinking otherwaies) I am induced to beleeue, by reason of the faithfull report that men sound of religion, haue made according to their sight thereof, I think if so be, I say these may be the respectes, whereupon the *Papistes* may haue that power. CHRIST gaue a commission and power to his Apostles to cast out Deuilles, which they according thereunto put in execution: The rules he bad them obserue in that action, was fasting and praier: & the action it selfe to be done in his name. This power of theirs proceeded not then of anie vertue in them, but onely in him who directed them. As was clearly proued by *Iudas* his hauing as greate power in that commission, as anie of the reste. It is easie then to be vnderstand that the casting out of Deuilles, is by the vertue of fasting and prayer, and in-calling of the name of God, suppose manie imperfectiones be in the person that is the instrumente,

Mat. 7.

as CHRIST him selfe teacheth vs of the power that false Prophets sall haue to caste out Devils. It is no wounder then, these respects of this action being considered, that it may be possible to the

DAEMONOLOGIE

Papistes, though erring in sundrie points of Religion to accomplish this, if they vse the right forme prescribed by CHRIST herein. For what the worse is that action that they erre in other thinges, more then their Baptisme is the worse that they erre in the other Sacrament, and haue eiked many vaine freittes to the Baptisme it selfe.

PHI. Surelie it is no little wonder that God should permit the bodies of anie of the faithfull to be so dishonoured, as to be a dwelling place to that vncleane spirite.

EPI. There is it which I told right now, would prooue and strengthen my argument of the deuils entring in the dead bodies of the faithfull. For if he is permitted to enter in their liuing bodies, euen when they are ioyned with the soule: how much more will God permit him to enter in their dead carions, which is no more man, but the filthie and corruptible caise of man. For as CHRIST sayth,

Mark. 7.

It is not any thing that enters within man that defiles him, but onely that which proccedes and commeth out of him.

Chapter 5

ARGVMENT.

The description of the fourth kinde of Spirites called the Phairie: *What is possible therein, and what is but illusiones. How far this Dialogue entreates of all these things, and to what end.*

PHILOMATHES.

Now I pray you come on to that fourth kinde of spirites.

EPI. That fourth kinde of spirites, which by the Gentiles was called *Diana*, and her wandring court, and amongst vs was called the *Phairie* (as I tould you) or our good neighboures, was one of the sortes of illusiones that was rifest in the time of *Papistrie*: for although it was holden odious to Prophesie by the deuill, yet whome these kinde of Spirites carryed awaie, and informed, they were thought to be sonsiest and of best life. To speake of the many vaine trattles founded vpon that illusion: How there was a King and Queene of *Phairie*, of such a iolly court & train as they had, how they had a teynd, & dutie, as it were, of all goods: how they naturallie rode and went, eate and drank, and did all other actiones like naturall men and women: I thinke it liker VIRGILS *Campi Elysij*, nor anie thing that ought to be beleeued by Christians, except in generall, that as I spake sundrie times before, the deuil illuded the senses of sundry simple creatures, in making them beleeue that they saw and harde such thinges as were nothing so indeed.

PHI. But how can it be then, that sundrie Witches haue gone to death with that confession, that they haue ben transported with the *Phairie* to such a hill, which opening, they went in, and there saw a faire Queene, who being now lighter, gaue them a stone that had sundrie vertues, which at sundrie times hath bene produced in judgement?

EPI. I say that, euen as I said before of that imaginar rauishing of the spirite foorth of the bodie. For may not the deuil object to their

DAEMONOLOGIE

fantasie, their senses being dulled, and as it were a sleepe, such hilles & houses within them, such glistering courts and traines, and whatsoeuer such like wherewith he pleaseth to delude them. And in the meane time their bodies being senselesse, to conuay in their hande any stone or such like thing, which he makes them to imagine to haue receiued in such a place.

PHI. But what say ye to their fore-telling the death of sundrie persones, whome they alleage to haue scene in these places? That is, a sooth-dreame (as they say) since they see it walking.

EPI. I thinke that either they haue not bene sharply inough examined, that gaue so blunt a reason for their Prophesie, or otherwaies, I thinke it likewise as possible that the Deuill may prophesie to them when he deceiues their imaginationes in that sorte, as well as when he plainely speakes vnto them at other times for their prophesying, is but by a kinde of vision, as it were, wherein he commonly counterfeits God among the Ethnicks, as I told you before.

PHI. I would know now whether these kindes of spirites may only appeare to Witches, or if they may also appeare to anie other.

EPI. They may do to both, to the innocent sort, either to affraie them, or to seeme to be a better sorte of folkes nor vncleane spirites are, and to the Witches, to be a cullour of safetie for them, that ignorant Magistrates may not punish them for it, as I told euen now. But as the one sorte, for being perforce troubled with them ought to be pittied, so ought the other sorte (who may bee discerned by their taking vppon them to Prophesie by them,) That sorte I say, ought as seuerely to be punished as any other Witches, and rather the more, that that they goe dissemblingly to woorke.

PHI. And what makes the spirites haue so different names from others.

EPI. Euen the knauerie of that same deuil; who as hee illudes the *Necromancers* with innumerable feyned names for him and his angels, as in special, making *Sathan, Beelzebub, & Lucifer*, to be three sundry spirites, where we finde the two former, but diuers names giuen to the Prince of all the rebelling angels by the Scripture. As by CHRIST, the Prince of all the Deuilles is called, *Beelzebub* in that place, which I alleaged against the power of any hereticques to cast out Deuils. By IOHN in the Reuelation, the old tempter is called, *Sathan the Prince of all the euill angels*. And the last, to wit, *Lucifer*, is but by allegoric taken from *the day Starre* (so named in diuers places of the Scriptures) because of his excellencie (I meane the Prince of them) in his creation before his fall. Euen so I say he deceaues the Witches, by attributing to himselfe diuers names: as if euery diuers shape that he trans formes himselfe in, were a diuers kinde of spirit.

PHI. But I haue hard many moe strange tales of this *Phairie*, nor ye haue yet told me.

EPI. As well I do in that, as I did in all the rest of my discourse. For because the ground of this conference of ours, proceeded of your speering at me at our meeting, if there was such a thing as Witches or spirites: And if they had any power: I therefore haue framed my whole discours, only to proue that such things are and may be, by such number of examples as I show to be possible by reason: & keepes me from dipping any further in playing the part of a Dictionarie, to tell what euer I haue read or harde in that purpose, which both would exceede fayth, and rather would seeme to teach such vnlawfull artes, nor to disallow and condemne them, as it is the duetie of all Christians to do.

Chapter 6

ARGVMENT.

Of the tryall and punishment of Witches. What sorte of accusation ought to be admitted against them. What is the cause of the increasing so far of their number in this age.

PHILOMATHES.

Then to make an ende of our conference, since I see it drawes late, what forme of punishment thinke ye merites these *Magicians* and Witches? For I see that ye account them to be all alike guiltie?

EPI. They ought to be put to death according to the Law of God, the ciuill and imperial law, and municipall law of all Christian nations.

PHI. But what kinde of death I pray you?

EPI. It is commonly vsed by fire, but that is an indifferent thing to be vsed in euery cuntrie, according to the Law or custome thereof.

PHI. But ought no sexe, age nor ranck to be exempted?

EPI. None at al (being so vsed by the lawful Magistrat) for it is the highest poynt of Idolatrie, wherein no exception is admitted by the law of God.

PHI. Then bairnes may not be spared?

EPI. Yea, not a haire the lesse of my conclusion. For they are not that capable of reason as to practise such thinges. And for any being in company and not reueiling thereof, their lesse and ignorant age will no doubt excuse them.

PHI. I see ye condemne them all that are of the counsell of such craftes.

EPI. No doubt, for as I said, speaking of *Magie*, the consulters, trusters in, ouer-seers, interteiners or sturrers vp of these craftes-folkes, are equallie guiltie with themselues that are the practisers.

PHI. Whether may the Prince then, or supreame Magistrate, spare or ouer-see any that are guiltie of that craft? vpon som great respects knowen to him?

EPI. The Prince or Magistrate for further tryals cause, may continue the punishing of them such a certaine space as he thinkes conuenient: But in the end to spare the life, and not to strike when God bids strike, and so seuerelie punish in so odious a fault & treason against God, it is not only vnlawful, but doubtlesse no lesse sinne in that Magistrate, nor it was in SAVLES sparing of AGAG. And so comparable

1. *Sam.* 15.

to the sin of Witch-craft it selfe, as SAMVELL alleaged at that time.

PHI. Surely then, I think since this crime ought to be so seuerely punished. Judges ought to beware to condemne any, but such as they are sure are guiltie, neither should the clattering reporte of a carling serue in so weightie a case.

EPI. Iudges ought indeede to beware whome they condemne: For it is as great a crime

Pro. 17.

(as SALOMON sayeth,) *To condemne the innocent, as to let the guiltie escape free*; neither ought the report of any one infamous person, be admitted for a sufficient proofe, which can stand of no law.

PHI. And what may a number then of guilty persons confessions, woork against one that is accused?

EPI. The assise must serue for interpretour of our law in that respect. But in my opinion, since in a mater of treason against the Prince, barnes or wiues, or neuer so diffamed persons, may of our law serue for sufficient witnesses and proofes. I thinke surely that by a far greater reason, such witnesses may be sufficient in matters of high treason against God: For who but Witches can be prooues, and so witnesses of the doings of Witches.

PHI. Indeed, I trow they wil be loath to put any honest man vpon their counsell. But what if they accuse folke to haue bene present at their Imaginar conuentiones in the spirite, when their bodies lyes sencelesse, as ye haue said.

EPI. I think they are not a haire the lesse guiltie: For the Deuill durst neuer haue borrowed their shaddow or similitude to that turne, if their consent had not bene at it: And the consent in these turnes is death of the law.

DAEMONOLOGIE

PHI. Then SAMVEL was a Witch: For the Deuill resembled his shape, and played his person in giuing response to SAVLE.

EPI. SAMVEL was dead aswell before that; and so none coulde slander him with medling in that vnlawfull arte. For the cause why, as I take it, that God will not permit Sathan to vse the shapes or similitudes of any innocent persones at such vnlawful times, is that God wil not permit that any innocent persons shalbe slandered with that vile defection: for then the deuil would find waies anew, to calumniate the best. And this we haue in proofe by them that are carryed with the *Phairie*, who neuer see the shaddowes of any in that courte, but of them that thereafter are tryed to haue bene brethren and sisters of that craft. And this was likewise proued by the confession of a young Lasse, troubled with spirites, laide on her by Witchcraft. That although shee saw the shapes of diuerse men & women troubling her, and naming the persons whom these shaddowes represents: yet neuer one of them are found to be innocent, but al clearely tried to be most guilty, & the most part of them confessing the same. And besides that, I think it hath ben seldome harde tell of, that any whome persones guiltie of that crime accused, as hauing knowen them to be their marrowes by eye-sight, and not by hear-say, but such as were so accused of Witch-craft, could not be clearely tryed vpon them, were at the least publickly knowen to be of a very euil life & reputation: so iealous is God I say, of the fame of them that are innocent in such causes. And besides that; there are two other good helpes that may be vsed for their trial: the one is the finding of their marke, and the trying the insensiblenes thereof. The other is their fleeting on the water: for as in a secret murther, if the deade carcase be at any time thereafter handled by the murtherer, it wil gush out of bloud, as if the blud wer crying to the heauen for reuenge of the murtherer, God hauing appoynted that secret super-naturall signe, for tryall of that secrete vnnaturall crime, so it appeares that God hath appoynted (for a super-naturall signe of the monstruous impietie of the Witches) that the water shal refuse to receiue them in her bosom, that haue shaken off them the sacred Water of Baptisme, and wilfullie refused the benefite thereof: No not so much as their eyes are able to shed teares (thretten and torture them as ye please) while first they repent (God not permitting them to dissemble their obstinacie in so horrible a crime) albeit the women kinde especially, be able other-waies to shed teares at euery light occasion when they will, yea, although it were dissemblingly like the *Crocodiles*.

PHI. Well, wee haue made this conference to last as long as leasure would permit: And to conclude then, since I am to take my leaue of you, I pray God to purge this Cuntrie of these diuellishe practises: for they were neuer so rife in these partes, as they are now.

EPI. I pray God that so be to. But the causes ar ouer manifest, that makes them to be so rife. For the greate wickednesse of the people on the one parte, procures this horrible defection, whereby God justlie punisheth sinne, by a greater iniquitie. And on the other part, the consummation of the worlde, and our deliuerance drawing neare,

Reuel. 12.

makes Sathan to rage the more in his instruments, knowing his kingdome to be so neare an ende. And so fare-well for this time.

FINIS.

Newes from Scotland

Declaring the Damnable *life and death of Doctor Fian, a* notable Sorcerer, who was burned at Edenbrough in Ianuary last. 1591.

Which Doctor was regester to the Diuell that sundry times preached at North Barrick Kirke, to a number of notorious Witches.

With the true examinations of the saide Doctor and Witches, as they vttered them in the presence *of the Scottish King.*

Discouering how they pretended *to bewitch and drowne his Maiestie in the Sea* comming from Denmark with such *other wonderfull matters as the like hath not been heard of at any time.*

Published according to the Scottish Coppie.
AT   LONDON
Printed for William *Wright.*

To the Reader

      The Manifolde vntruthes which is spread abroade, concerning the detestable actions and apprehension of those Witches wherof this Historye following truely entreateth, hath caused me to publish the same in print: and the rather for that sundrie written Copies are lately dispersed therof, containing, that the said witches were first discouered, by meanes of a poore Pedler trauailing to the towne of *Trenent*, and that by a wonderfull manner he was in a moment conuayed at midnight, from *Scotland* to *Burdeux* in *Fraunce* (beeing places of no small distance between) into a Marchants Seller there, & after, being sent from *Burdeux* into *Scotland* by certaine Scottish Marchants to the Kinges Maiestie, that he discouered those Witches and was the cause of their apprehension: with a number of matters miraculous and incredible: All which in truthe are moste false. Neuertheles to satisfie a number of honest mindes, who are desirous to be enformed of the veritie and trueth of their confessions, which for certaintie is more stranger then the common reporte runneth, and yet with more trueth I haue undertaken to publish this short Treatise, which declareth the true discourse of all that hath hapned, & aswell what was pretended by those wicked and detestable Witches against the Kinges Maiestie, as also by what meanes they wrought the same.

      All which examinations (gentle Reader) I haue heere truelye published, as they were taken and uttered in the presence of the Kings Maiestie, praying thee to accept it for veritie, the same beeing so true as cannot be reproued.

Discourse

A true discourse, of the apprehension of sundrye Witches lately taken in Scotland, some are executed, and some are yet imprisoned.

With a particuler recitall of their examinations, taken in the presence of the Kinges Maiestie.

God by his omnipotent power, hath at al times and daily doth take such care, and is so vigillant, for the weale and preseruation of his owne, that thereby he disapointeth the wicked practises and euil intents of all such as by any meanes whatsoeuer, seeke indirectly to conspire any thing contrary to his holy will: yea and by the same power, he hath lately ouerthrown and hindered the intentions and wicked dealinges of a great number of vngodly creatures, no better then Diuels: who suffering themselues to be allured and inticed by the Diuell whom they serued, and to whome they were priuatelye sworne: entered into the detestable Art of witchcraft, which they studied and practised so long time, that in the end they had seduced by their sorcery a number of other to be as bad as themselues: dwelling in the boundes of *Lowthian*, which is a principall shire or parte of *Scotland*, where the Kings Maiestie vseth to make his cheefest residence or abode: and to the end that their detestable wickednes which they priuilye had pretended against the Kings Maiestie, the Common-weale of that Country, with the Nobilitie and subjects of the same, should come to light: God of his vnspeakeable goodnes did reueale and lay it open in very strange sorte, therby to make knowne vnto the worlde, that there actions were contrarye to the lawe of God, and the naturall affection which we ought generallye to beare one to another: the manner of the reuealing wherof was as followeth.

Within the towne of *Trenent* in the Kingdome of *Scotland*, there dwelleth one *Dauid Seaton*, who being deputie Bailiffe in the saide Towne, had a maide seruant called *Geillis Duncane*, who vsed secretly to be absent and to lye foorth of her Maisters house euery other night: this *Geillis Duncane* took in hand to help all such as were troubled or greeued with any kinde of sicknes or infirmitie: and in short space did perfourme manye matters most miraculous, which thinges forasmuch as she began to doe them vpon a sodaine, hauing neuer doon the like before, made her Maister and others to be in great admiracion, and wondred thereat: by meanes wherof the saide *Dauid Seaton* had his maide in some great suspition, that she did not those things by naturall and lawfull wayes, but rather supposed it to be doone by some extraordinary and vnlawfull meanes.

Whervpon, her Maister began to growe very inquisitiue, and examined her which way and by what meanes she were able to perfourme matters of so great importance: whereat she gaue him no answere, neuerthelesse, her Maister to the intent that he might the better trye and finde out the trueth of the same, did with the helpe of others, torment her with the torture of the Pilliwinckes vpon her fingers, which is a greeuous torture, and binding or wrinching her head with a corde or roape, which is a most cruell torment also, yet would she not confesse any thing, whereupon they suspecting that she had beene marked by the Diuell (as commonly witches are) made dilligent search about her, and found the enemies marke to be in her fore crag or foreparte of her throate: which being found, she confessed that all her dooings was doone by the wicked allurements and inticements of the Diuell, and that she did them by witchcraft.

DAEMONOLOGIE

After this her confession, she was committed to prison, where she continued for a season, where immediatly she accused these persons following to be notorious witches, and caused them foorthwith to be apprehended one after an other, vidz. *Agnis Sampson* the eldest Witch of them al, dwelling in Haddington, *Agnes Tompson* of Edenbrough, Doctor *Fian, alias Iohn Cunningham*, maister of the Schoole at Saltpans in Lowthian, of whose life and strange actes, you shall heare more largely in the ende of this discourse: these were by the saide *Geillis Duncane* accused, as also *George Motts* wife dwelling in Saltpans, *Robert Griersonn* skipper, and *Iennit Bandilandis*, with the Porters wife of Seaton, the Smith at the brigge Hallis with innumerable others in that partes, and dwelling in those bounds aforesaide: of whom some are alreadye executed, the rest remaine in prison, to receiue the doome of Iudgement at the Kings maiesties will and pleasure.

The said *Geillis Duncane* also caused *Ewphame Meealrean* to be apprehended, who conspired and perfourmed the death of her Godfather, and who vsed her art vpon a gentleman being one of the Lords and Iustices of the Session, for bearing good will to her Daughter: she also caused to be apprehended one *Barbara Naper*, for bewitching to death *Archibalde*, last Earle of Angus, who languished to death by witchcraft and yet the same was not suspected, but that he died of so strange a disease, as the Phisition knew not how to cure or remedy the same: but of all other the saide witches, these two last before recited, were reputed for as ciuill honest women as any that dwelled within the Citie of Edenbrough, before they were apprehended. Many other besides were taken dwelling in Lieth, who are detayned in prison, vntill his Maiesties further will and pleasure be known: of whose wicked dooings you shall particularly heare, which was as followeth.

This aforeaside *Agnis Sampson* which was the elder Witch, was taken and brought to Haliciud house before the Kings Maiestie and sundry other of the nobility of Scotland, where she was straitly examined, but all the perswasions which the Kings maiestie vsed to her with the rest of his counsell, might not prouoke or induce her to confesse any thing, but stood stiffely in the deniall of all that was laide to her charge: whervpon they caused her to be conueied awaye to prison, there to receiue such torture as hath been lately prouided for witches in that country: and forasmuch as by due examination of witchcraft and witches in Scotland, it hath latelye beene found that the Deuill dooth generallye marke them with a priuie marke, by reason the Witches haue confessed themselues, that the Diuell dooth lick them with his tung in some priuy part of their bodie, before hee dooth receiue them to be his seruants, which marke commonly is giuen them vnder the haire in some part of their bodye, wherby it may not easily be found out or scene, although they be searched: and generally so long as the marke is not seene to those which search them, so long the parties that hath the marke will neuer confesse any thing. Therfore by special commaundement this *Agnis Sampson* had all her haire shauen of, in each parte of her bodie, and her head thrawen with a rope according to the custome of that Countrye, beeing a paine most greeuous, which she continued almost an hower, during which time she would not confesse any thing vntill the Diuels marke was found vpon her priuities, then she immediatlye confessed whatsoeuer was demaunded of her, and iustifying those persons aforesaid to be notorious witches.

DAEMONOLOGIE

    Item, the saide *Agnis Tompson* was after brought againe before the Kings Maiestie and his Counsell, and being examined of the meetings and detestable dealings of those witches, she confessed that vpon the night of *Allhollon* Euen last, she was accompanied aswell with the persons aforesaide, as also with a great many other witches, to the number of two hundreth: and that all they together went by Sea each one in a Riddle or Ciue, and went in the same very substantially with Flaggons of wine making merrie and drinking by the waye in the same Riddles or Ciues, to the Kerke of North Barrick in Lowthian, and that after they had landed, tooke handes on the land and daunced this reill or short daunce, singing all with one voice. *Commer goe ye before, commer goe ye,*
    If ye will not goe before, commer let me.
    At which time she confessed, that this *Geilles Duncane* did goe before them playing this reill or daunce vpon a small Trump, called a Iewes Trump, vntill they entred into the Kerk of north Barrick.
    These confessions made the King in a woderful admiration, and sent for the said *Geillis Duncane*, who vpon the like Trump did playe the said daunce before the Kings Maiestie, who in respect of the strangenes of these matters, tooke great delight to bee present at their examinations.
    Item, the said *Agnis Tompson* confessed that the Diuell being then at North Barrick Kerke attending their comming in the habit or likenes of a man, and seeing that they tarried ouer long, he at their comming enioyned them all to a pennance, which was, that they should kisse his Buttockes, in signe of duetye to him: which being put ouer the Pulpit barre, euerye one did as he had enioyned them: and hauing made his vngodly exhortations, wherein he did greatlye enveighe against the King of Scotland, he receiued their oathes for their good and true seruice towards him, and departed: which doone, they returned to Sea, and so home againe. At which time the witches demaunded of the Diuel why he did beare such hatred to the King, who answered, by reason the King is the greatest enemy he hath in the worlde: all which their confessions and depositions are still extant vpon record.
    Item, the saide *Agnis Sampson* confessed before the Kings Maiestie sundrye thinges which were so miraculous and strange, as that his Maiestie saide they were all extreame lyars, wherat she answered, she would not wishe his Maiestie to suppose her woords to be false, but rather to beleeue them, in that she would discouer such matter vnto him as his maiestie should not any way doubt off.

And therupon taking his Maiestie a little aside, she declared vnto him the verye woordes which passed betweene the Kings Maiestie and his Queene at Vpslo in Norway the first night of their mariage, with their answere eache to other: whereat the Kinges Maiestie wondered greatlye, and swore by the liuing God, that he beleeued that all the Diuels in hell could not haue discouered the same: acknowledging her woords to be most true, and therefore gaue the more credit to the rest which is before declared.

Touching this *Agnis Tompson*, she is the onlye woman, who by the Diuels perswasion should haue entended and put in execution the Kings Maiesties death in this manner.

She confessed that she tooke a blacke Toade, and did hang the same vp by the heeles, three daies, and collected and gathered the venome as it dropped and fell from it in an Oister shell, and kept the same venome close couered, vntill she should obtaine any parte or peece of foule linnen cloth, that had appertained to the Kings Maiestie, as shirt, handkercher, napkin or any other thing which she practised to obtaine by meanes of one *Iohn Kers*, who being attendant in his Maiesties Chamber, desired him for olde acquaintance betweene them, to helpe her to one or a peece of such a cloth as is aforesaide, which thing the said *Iohn Kers* denyed to helpe her too, saying he could not help her too it.

And the said *Agnis Tompson* by her depositions since her apprehension saith, that if she had obtained any one peece of linnen cloth which the King had worne and fouled, she had bewitched him to death, and put him to such extraordinary paines, as if he had beene lying vpon sharp thornes and endes of Needles.

Moreouer she confessed that at the time when his Maiestie was in Denmarke, she being accompanied with the parties before specially named, tooke a Cat and christened it, and afterward bound to each parte of that Cat, the cheefest partes of a dead man, and seuerall ioynts of his bodie, and that in the night following the saide Cat was conueied into the midst of the sea by all these witches sayling in their riddles or Ciues as is aforesaide, and so left the saide Cat right before the Towne of Lieth in Scotland: this doone, there did arise such a tempest in the Sea, as a greater hath not beene seene: which tempest was the cause of the perrishing of a Boate or vessell comming ouer from the towne of Brunt Iland to the towne of Lieth, wherein was sundrye Iewelles and riche giftes, which should haue been presented to the now Queen of Scotland, at her Maiesties comming to Lieth.

Againe it is confessed, that the said christened Cat was the cause that the Kinges Maiesties Ship at his comming foorth of Denmarke, had a contrary winde to the rest of his Ships, then being in his companye, which thing was most strange and true, as the Kings Maiestie acknowledgeth, for when the rest of the Shippes had a faire and good winde, then was the winde contrarye and altogither against his Maiestie: and further the saide witche declared, that his Maiestie had neuer come safelye from the Sea, if his faith had not preuailed aboue their ententions.

Moreouer the said Witches being demaunded how the Diuell would vse them when he was in their company, they confessed that when the Diuell did receiue them for his seruants, and that they had vowed themselues vnto him, then he would Carnallye vse them, albeit to their little pleasure, in respect of his colde nature: and would doo the like at sundry other times.

As touching the aforesaide Doctor *Fian, alias Iohn Cunningham,* the examination of his actes since his apprehension, declareth the great subtiltye of the diuell, and therfore maketh thinges to appeere the more miraculous: for being apprehended by the accusation of the saide *Geillis Duncane* aforesaide, who confessed he was their Regester, and that there was not one man suffered to come to the Diuels readinges but onlye he: the saide Doctor was taken and imprisoned, and vsed with the accustomed paine, prouided for those offences, inflicted vpon the rest as is aforesaide.

First by thrawing of his head with a roape, wherat he would confesse nothing.

Secondly, he was perswaded by faire means to confesse his follies, but that would preuaile as little.

Lastly he was put to the most seuere and cruell paine in the world, called the bootes, who after he had receiued three strokes, being enquired if he would confesse his damnable acts and wicked life, his tung would not serue him to speak, in respect wherof the rest of the witches willed to search his tung, vnder which was found two pinnes thrust vp into the head, whereupon the witches did laye, *Now is the Charme stinted*, and shewed that those charmed Pinnes were the cause he could not confesse any thing: then was he immediatly released of the bootes, brought before the King, his confession was taken, and his owne hand willingly set ther-vnto, which contained as followeth.

First, that at the generall meetinges of those witches, hee was alwayes preasent: that he was Clarke to all those that were in subiection to the Diuels seruice, bearing the name of witches, that alwaye he did take their othes for their true seruice to the Diuell, and that he wrot for them such matters as the Diuell still pleased to commaund him.

Item, he confessed that by his witchcrafte he did bewitch a Gentleman dwelling neere to the Saltpans, where the said Doctor kept Schoole, onely for being enamoured of a Gentlewoman whome he loued himselfe: by meanes of which his Sorcerye, witchcraft and diuelish practises, he caused the said Gentleman that once in xxiiij. howres he fell into a lunacie and madnes, and so continued one whole hower together, and for the veritie of the same, he caused the Gentleman to be brought before the Kinges Maiestie, which was vpon the xxiiij. day of December last, and being in his Maiesties Chamber, suddenly he gaue a great scritch and fell into a madnes, sometime bending himselfe, and sometime capring so directly vp, that his head did touch the seeling of the Chamber, to the great admiration of his Maiestie and others then present: so that all the Gentlemen in the Chamber were not able to holde him, vntill they called in more helpe, who together bound him hand and foot: and suffering the said gentleman to lye still vntill his furye were past, he within an hower came againe to himselfe, when being demaunded of the Kings Maiestie what he saw or did all that while, answered that he had been in a sound sleepe.

Item the said Doctor did also confesse that he had vsed means sundry times to obtain his purpose and wicked intent of the same Gentlewoman, and seeing himselfe disapointed of his intention, he determined by all waies he might to obtaine the same, trusting by coniuring, witchcraft and Sorcery to obtaine it in this manner.

It happened this gentlewoman being vnmaried, had a brother who went to schoole with the said Doctor, and calling his Scholler to him, demaunded if he did lye with his sister, who answered he did, by meanes wherof he thought to obtaine his purpose, and therefore secretlye promised to teach him without stripes, so he would obtain for him three haires of his sisters priuities, at such time as he should spye best occasion for it: which the youth promised faithfullye to perfourme, and vowed speedily to put it in practise, taking a peece of coniured paper of his maister to lappe them in when he had gotten them: and therevpon the boye practised nightlye to obtaine his maisters purpose, especially when his sister was a sleepe.

But God who knoweth the secrets of all harts, and reuealeth all wicked and vngodlye practises, would not suffer the intents of this diuilish Doctor to come to that purpose which he supposed it would,

DAEMONOLOGIE

and therefore to declare that he was heauilye offended with his wicked entent, did so woorke by the Gentlewomans owne meanes, that in the ende the same was discouered and brought to light: for she being one night a sleepe, and her brother in bed with her, suddenlye cryed out to her mother, declaring that her Brother would not suffer her to sleepe, wherevpon her mother hauing a quick capacitie, did vehemently suspect Doctor *Fians* entention, by reason she was a witche of her selfe, and therefore presently arose, and was very inquisitiue of the boy to vnderstand his intent, and the better to know the same, did beat him with sundry stripes, wherby he discouered the trueth vnto her. The Mother therefore being well practised in witchcrafte, did thinke it most conuenient to meete with the Doctor in his owne Arte, and therevpon tooke the paper from the boy, wherein hee should haue put the same haires, and went to a young Heyfer which neuer had borne Calfe nor gone to the Bull, and with a paire of sheeres, clipped off three haires from the vdder of the Cow, and wrapt them in the same paper, which she againe deliuered to the boy, then willing him to giue the same to his saide Maister, which he immediatly did.

The Schoolemaister so soone as he had receiued them, thinking them indeede to bee the Maides haires, went straight and wrought his arte vpon them: But the Doctor had no sooner doone his intent to them, but presentlye the Hayfer or Cow whose haires they were indeed, came vnto the doore of the Church wherein the Schoolemaister was, into the which the Hayfer went, and made towards the Schoolemaister, leaping and dauncing vpon him, and following him foorth of the church and to what place so euer he went, to the great admiration of all the townes men of Saltpans, and many other who did beholde the same.

The reporte whereof made all men imagine that hee did woorke it by the Diuell, without whom it could neuer haue beene so sufficientlye effected: and thervpon, the name of the said Doctor *Fian* (who was but a very yong man) began to grow so common among the people of Scotland, that he was secretlye nominated for a notable Cuniurer.

JAMES I OF ENGLAND

All which although in the beginning he denied, and would not confesse, yet hauing felt the pain of the bootes (and the charme stinted, as aforesayd) he confessed all the aforesaid to be most true, without producing anie witnesses to iustifie the same, & thervpon before the kings maiesty he subscribed the sayd confessions with his owne hande, which for truth remaineth vpon record in *Scotland*.

After that the depositions and examinations of the sayd doctor *Fian Alias Cuningham* was taken, as alreadie is declared, with his owne hand willingly set therevnto, hee was by the master of the prison committed to ward, and appointed to a chamber by himselfe, where forsaking his wicked wayes, acknowledging his most vngodly lyfe, shewing that he had too much folowed the allurements and entisements of Sathan, and fondly practised his conclusions by coniuring, witchcraft, inchantment, sorcerie, and such like, hee renounced the deuill and all his wicked workes, vowed to leade the life of a Christian, and seemed newly connected towards God.

DAEMONOLOGIE

The morrow after vpon conference had with him, he granted that the deuill had appeared vnto him in the night before, appareled all in blacke, with a white wand in his hande, and that the deuill demaunded of him if hee would continue his faithfull seruice, according to his first oath and promise made to that effect. Whome (as hee then sayd) he vtterly renounced to his face, and sayde vnto him in this manner, *Auoide Satan, auoide*, for I haue listned too much vnto thee, and by the same thou hast vndone mee, in respect whereof I vtterly forsake thee. To whome the deuill answered, *That once ere thou die thou shall bee mine*. And with that (as he sayde) the deuill brake the white wande, and immediatly vanished foorth of his sight.

Thus all the daie this Doctor Fian continued verie solitarie, and seemed to haue care of his owne soule, and would call vppon God, shewing himselfe penitent for his wicked life, neuerthelesse the same night hee founde such meanes, that hee stole the key of the prison doore and chamber in the which he was, which in the night hee opened and fled awaie to the Salt pans, where hee was always resident, and first apprehended. Of whose sodaine departure when the Kings maiestie had intelligence, hee presently commanded diligent inquirie to bee made for his apprehension, and for the better effecting thereof, hee sent publike proclamations into all partes of his lande to the same effect. By meanes of whose hot and harde pursuite, he was agayn taken and brought to prison, and then being called before the kings highnes, hee was reexamined as well touching his departure, as also touching all that had before happened. But this Doctor, notwithstanding that his owne confession appeareth remaining in recorde vnder his owne hande writing, and the same therevnto fixed in the presence of the Kings maiestie and sundrie of his Councell, yet did hee vtterly denie the same.

Wherevpon the kinges maiestie perceiuing his stubbourne wilfulnesse, concerned and imagined that in the time of his absence hee had entered into newe conference and league with the deuill his master, and that hee had beene agayne newly marked, for the which hee was narrowly searched, but it coulde not in anie wise bee founde, yet for more tryall of him to make him confesse, hee was commaunded to haue a most straunge torment which was done in this manner following.

His nailes vpon all his fingers were riuen and pulled off with an instrument called in Scottish a *Turkas*, which in England wee call a payre of pincers, and vnder euerie nayle there was thrust in two needels ouer euen up to the heads. At all which tormentes notwithstanding the Doctor neuer shronke anie whit, neither woulde he then confesse it the sooner for all the tortures inflicted vpon him.

Then was hee with all conuenient speed, by commandement, conuaied againe to the torment of the bootes, wherein hee continued a long time, and did abide so many blowes in them, that his legges were crushte and beaten togeather as small as might bee, and the bones and flesh so brused, that the bloud and marrowe spouted forth in great abundance, whereby they were made unseruiceable for euer. And notwithstanding al these grieuous paines and cruell torments hee would not confesse anie thing, so deepely had the deuill entered into his heart, that hee vtterly denied all that which he had before auouched, and woulde saie nothing therevnto but this, that what hee had done and sayde before, was onely done and sayde for feare of paynes which he had endured.

Upon great consideration therefore taken by the Kings maiestie and his Councell, as well for the due execution of iustice vppon such detestable malefactors, as also for example sake, to remayne a terrour to all others heereafter, that shall attempt to deale in the lyke wicked and vngodlye actions, as witchcraft, sorcery, cuniuration, & such lyke, the sayde Doctor *Fian* was soone after araigned, condemned, and adiudged by the law to die, and then to bee burned according to the lawe of that lande, prouided in that behalfe. Wherevpon hee was put into a carte, and beeing first strangled, hee was immediatly put into a great fire, being readie prouided for that purpose, and there burned in the Castle hill of *Edenbrough* on a saterdaie in the ende of Ianuarie last past. 1591. The rest of the witches which are not yet executed, remayne in prison till farther triall, and knowledge of his maiesties pleasure.

This strange discourse before recited, may perhaps giue some occasion of doubt to such as shall happen to reade the same, and thereby coniecture that the Kings maiestie would not hazarde himselfe in the presence of such notorious witches, least therby might haue insued great danger to his person and the generall state of the land, which thing in truth might wel haue bene feared. But to answer generally to such, let this suffice: that first it is well knowen that the King is the child & seruant of God, and they but seruants to the deuil, hee is the Lords annointed, and they but vesselles of Gods wrath: he is a true Christian, and trusteth in God, they worse than Infidels, for they onely trust in the deuill, who daily serue them, till he haue brought them to vtter destruction. But heereby it seemeth that his Highnesse carted a magnanimious and undanted mind, not feared with their

DAEMONOLOGIE

inchantmentes, but resolute in this, that so long as God is with him, hee feareth not who is against him. And trulie the whole scope of this treatise dooth so plainely laie open the wonderfull prouidence of the Almightie, that if he had not bene defended by his omnipotencie and power, his Highnes had neuer returned aliue in his voiage frõ Denmarke, so that there is no doult but God woulde as well defend him on the land as on the sea, where they pretended their damnable practise.

Also from Benediction Books ...
Wandering Between Two Worlds: Essays on Faith and Art
Anita Mathias
Benediction Books, 2007
152 pages
ISBN: 0955373700

Available from www.amazon.com, www.amazon.co.uk

In these wide-ranging lyrical essays, Anita Mathias writes, in lush, lovely prose, of her naughty Catholic childhood in Jamshedpur, India; her large, eccentric family in Mangalore, a sea-coast town converted by the Portuguese in the sixteenth century; her rebellion and atheism as a teenager in her Himalayan boarding school, run by German missionary nuns, St. Mary's Convent, Nainital; and her abrupt religious conversion after which she entered Mother Teresa's convent in Calcutta as a novice. Later rich, elegant essays explore the dualities of her life as a writer, mother, and Christian in the United States-- Domesticity and Art, Writing and Prayer, and the experience of being "an alien and stranger" as an immigrant in America, sensing the need for roots.

About the Author

Anita Mathias is the author of *Wandering Between Two Worlds: Essays on Faith and Art.* She has a B.A. and M.A. in English from Somerville College, Oxford University, and an M.A. in Creative Writing from the Ohio State University, USA. Anita won a National Endowment of the Arts fellowship in Creative Nonfiction in 1997. She lives in Oxford, England with her husband, Roy, and her daughters, Zoe and Irene.

Visit Anita's website
   http://www.anitamathias.com,
and Anita's blog
   http://dreamingbeneaththespires.blogspot.com, (Dreaming Beneath the Spires).

The Church That Had Too Much
Anita Mathias
Benediction Books, 2010
52 pages
ISBN: 9781849026567

Available from www.amazon.com, www.amazon.co.uk

The Church That Had Too Much was very well-intentioned. She wanted to love God, she wanted to love people, but she was both hampered by her muchness and the abundance of her possessions, and beset by ambition, power struggles and snobbery. Read about the surprising way The Church That Had Too Much began to resolve her problems in this deceptively simple and enchanting fable.

About the Author

Anita Mathias is the author of *Wandering Between Two Worlds: Essays on Faith and Art.* She has a B.A. and M.A. in English from Somerville College, Oxford University, and an M.A. in Creative Writing from the Ohio State University, USA. Anita won a National Endowment of the Arts fellowship in Creative Nonfiction in 1997. She lives in Oxford, England with her husband, Roy, and her daughters, Zoe and Irene.

Visit Anita's website
 http://www.anitamathias.com,
and Anita's blog
 http://dreamingbeneaththespires.blogspot.com (Dreaming Beneath the Spires).

www.ingramcontent.com/pod-product-compliance
Lightning Source LLC
LaVergne TN
LVHW041205250326
834689LV00001BA/16

9781849023047